HOLDING COMMON GROUND

HOLDING COMMON GROUND

THE INDIVIDUAL AND PUBLIC LANDS IN THE AMERICAN WEST

Edited by Paul Lindholdt and Derrick Knowles

Eastern Washington University Press
Spokane, Washington

Copyright © 2005 by Paul Lindholdt and Derrick Knowles.

All rights reserved.

Printed in the United States of America.

Book and cover design by John Laursen at Press-22, Portland, Oregon.

No part of this book may be reproduced or transmitted in any form or by any means, electronic or mechanical, including photocopying, recording, or by any information storage and retrieval system, without the permission in writing from the copyright owner.

ISBN 1-59766-000-0

Library of Congress Cataloging-in-Publication Data

    Holding common ground : the individual and public lands in the American West / edited by Paul Lindholdt and Derrick Knowles.

    p. cm.

    1. Environmentalism—West (U.S.). 2. Environmental protection—West (U.S.)—Citizen participation. I. Lindholdt, Paul J. II. Knowles, Derrick.

    GE198.W47H65 2005

    333.72'0978—dc22

                                 2004030976

Acknowledgments

Two of the following essays have been published previously, and we thank the authors and the magazines for the generous permissions to reprint them here. "Amazing Grace," by Kathleen Dean Moore, originally appeared in *Audubon*, March-April 2001. "Finding the Forest," by Michael Branch, originally appeared in *Orion Afield*, 195 Main Street, Great Barrington, Massachusetts 01230; 888-909-6568; www.oriononline.org ($30/year eight issues, four each of *Orion* and *Orion Afield*).

For permission to publish his photograph of the Palos Verdes Peninsula, Stacy Warren thanks Bruce Perry.

Larry N. Olson, the photographer of *Oregon Rivers* (Westcliffe Publishers, 1997), created the cover photograph of the Joseph Creek watershed, high in the Blue Mountains of northeastern Oregon. Copyright © 1997 by Larry N. Olson.

Contents

Foreword
Mitchell Thomashow
1

Hearing the Call
Paul Lindholdt
3

Section I
Awakenings

White Sands Missile Range
Sharman Apt Russell
11

The Middle Ground
Sherry Simpson
18

Northwest Passages
Laird Christensen
23

Going Wild
David Axelrod
28

Slate Mountain
Will Peterson
33

Full-Stomach Wilderness and the Suburban Esthetic
Harold Fromm
36

Rattlesnakes, Shipwrecks, and Terra Incognita
in Palos Verdes Peninsula
Stacy Warren
41

Dances with Bears
Jim Dwyer
48

Finding the Forest:
Citizen Activism in the Truckee River Watershed
Michael P. Branch
53

Right in My Back Yard, or Confessions of a RIMBY
T. Louise Freeman-Toole
59

Section II
Challenges

Wedded to the Cause
Robert Schnelle
67

The River Home
Amanda Gordon
71

Civil Disobedience and the Arctic National Wildlife Refuge
Carolyn Kremers
75

Buffalo Field Campaign
Dan Brister
79

Dear Mom and Dad—Send Food
Chuck Pezeshki
84

Looking Down from Hale-Bopp
Colin Chisholm
89

The Toll Road
Lee Schweninger
92

War Among the Saguaros
Bruce D. Eilerts
97

The Troubling Dawn
Estar Holmes
104

Clearwater Journal
William Johnson
109

Section III
Victories

Lost and Found

Terrell Dixon

117

Great Sand Dunes: The Shape of the Wind

Stephen Trimble

122

A Little Garden of Sand:
Bringing Back San Francisco's Native Dunes

Christine Colasurdo

127

Postcards from the Pleistocene:
Saving Hendrickson Canyon

Robert Michael Pyle

132

Long Canyon

Jerry Pavia

139

Turnaround for a Threatened Wilderness

Karen Tweedy-Holmes

144

Herons, Eagles, People, and Parks:
A Cautionary Tale from a British Columbia Gulf Island

Rebecca Raglon

148

Chaparral Gold: Point Mugu
Bill Weiler
153

In View of the Condor
Bradley John Monsma
160

Amazing Grace
Kathleen Dean Moore
165

Notes on Contributors
171

Foreword

Mitchell Thomashow

What motivates people to seek affiliation with landscape, habitats, and critters? Why do some people take such pleasure in observing nature that they dedicate their lives to cultivating this practice? What are the origins of such identification? How does it influence one's life decisions—choices about residence, profession, family, politics, and the routines and habits of daily work and play?

These questions are the Holy Grail of environmental education. We assume that the sources of deepest learning come from a person's engagement with the natural world. We wish to probe that engagement, hoping that engagement prompts concern, and that with concern comes care. Developing an ethic of care, in ourselves and in our students, is the moral imperative of good teaching and learning, as well as the root of ecological citizenship.

An instructive strategy for probing the depths of moral action is to ask people to describe experiences that reveal the layers of their ecological identity. Can you think of the moments in your life when your images of self and place integrated seamlessly? What did you do in those moments, and what were you thinking? Eventually, you string those moments together and they become the narrative of your life, providing you with a sense of purpose, a sense of belonging to something greater than yourself, a way to compose your ecological identity.

Holding Common Ground is a remarkable collection of these moments, strung together like beads on a string, several dozen episodes in which engagement, concern, and care sprout to the forefront of awareness. I savor the variety displayed here—heroic deeds privately taken, selfless conduct in anonymous settings, collective actions spontaneously emerged, strategic decisions carved from necessity. These tales, taken together, reveal the significance of small and incremental actions, inspired by place and community, lending meaning to the metaphor "grassroots." The power of

Holding Common Ground is the grace of reflective action. Here are writers, homemakers, policy wonks, and all manner of amateurs and professionals thinking carefully about the meaning of their lives, as their moral obligations unfold in the narratives of self and place. Surely we all have stories such as these. This collection inspires us to write more of them, to find them in classrooms and community centers, on talk radio or web chat rooms, while waiting in line in airports, or while discussing the weather with your neighbor. Let's elicit the narratives of ecological citizenship wherever they might emerge.

Abraham Joshua Heschel, the great Jewish philosopher, wrote that "indifference to the sublime wonder of living is the root cause of sin." A life spent observing nature in the context of self and place will be forever engaged, rarely indifferent. There are just too many landscapes to roam, habitats to explore, and creatures to love. There are too many stories to tell. How else do we work through the contradictions, ambiguities, and tensions of living in place? This collection reveals the inspiration contained in a small corner of one's life experience, how the sublime wonder of living is too big to ignore, too mysterious to explain, too precious to avoid. Read these essays, take heart in their uncommon ubiquity, and know that thousands more live in each and every one of us, filled with more gratitude and grace than you thought you had within you.

Hearing the Call
Paul Lindholdt

If people shape nature, natural places also can shape human character and culture. This theme anchors the essays in *Holding Common Ground* and demonstrates a trend in American letters—more narratives today are bending the literary genres by blending personal revelations with explorations of landscapes. The present essays, all of which unfold on public lands in the American West, including the West's waterways and stunning viewscapes, demonstrate the sincerity of passion and emotion that drives so many Americans to environmental activism.

The narratives in *Holding Common Ground* combine a geographical landscape—public lands in the American West—with more intimate narrative landscapes of self, family, and community. In his book *Taking Care*, William Kittredge, Old West rancher turned New West writer and activist, argues that "We live in stories. What we are is stories. We do things because of what is called character, and our character is formed by the stories we learn to live in." Stories, Kittredge points out, nourish and shape us. They offer maps that help us navigate our rapidly changing cultural geography, as contributor Stacey Warren learned in sharing details of her former home in coastal California. The personal stories in this book, as diverse as the landscapes that give rise to them, take root in three types of experience with nature. Sherry Simpson's and Lee Schweninger's narratives, for instance, represent pivotal *awakenings*—recognitions or epiphanies. William Johnson and Dan Brister typify the *challenges* or struggles that activists face. Rebecca Raglon and Stephen Trimble exemplify environmental *victories* or successes. Sharing such stories, as Terry Tempest Williams suggests in *Red*, her literary tribute to the redrock wilderness of southern Utah, helps us find our way back from divisive opinions towards conversation. "Story bypasses rhetoric and pierces the heart," she asserts. "Story offers a wash of images and emotion that returns us to our highest and deepest selves, where we remember what it means to be human, living in place with our neighbors." Sharing these stories can help us find common ground over divisive land-use issues by giving human faces to contested public lands. An understanding of what motivates Americans to protect those public lands from logging, grazing, mining, dams, and development may follow from these highly personal tales.

Wallace Stegner identified an "impassioned protectiveness" that citizen-activists feel for the western landscape. "The battlegrounds of the environmental movement," he asserted, "lie in the western public lands." And so we focused on the western United States, where public lands make up roughly half the total landmass, and on one Canadian province. Managed by a patchwork of under-funded agencies, public lands have become cultural and ecological flashpoints, as Chuck Pezeshki shows in his tale about the killing of bison outside Yellowstone. The authors of these essays represent a fair cross-section of people who draw inspiration from the lands in the American West.

In his novels and essays, Wallace Stegner characterized the West as a region of too much transience and too few emotional ties to place. This same refrain unites a number of essays here. In "Northwest Passages," Laird Christensen recalls, "If I stayed in one place more than a few months, the urge to be back on the road grew insistent—it was being between places that seemed to me most satisfying." Boom-and-bust economies, fleeting company towns, the disposition to cut and run—these are corporate patterns of mobility that humans in the American West have come to accept. Such patterns manifest in a forcibly transient citizenry (think hard-rock miners) whose conception of home is a new job in a new town for a few short months or years.

Despite the continuing transience in the West, many of the writers of these essays engage actively in the work of reconnecting with rivers, forests, and deserts. There is mystery, beauty, and healing in nature, as these writers attest. There are rewards in struggle and challenge, great joy in working for justice and change. Christine Colasurdo exhumes the Great Sand Dune on which much of San Francisco rests—exhumes it by constructing public gardens of native plants "tough enough to survive" the climate and the developments that entomb the dune. In wheat country, Louise Freeman-Toole works to restore larkspur, balsamroot, lupine, and fescue on the plateau called the Palouse along the Washington-Idaho border. Both women are artists of landscapes, scholars of native plants, activists of soils. To preserve when we can, restore when we must, to write landscapes alive for absent readers' eyes—such environmental aspirations complement one another. Subtle efforts are no less worthy and painstaking in their own way than the civil disobedience Carolyn Kremers advocates to save the Arctic National Wildlife Refuge, or the lobbying of the California legislature that Bill Weiler describes.

Rhapsody and attachment characterize classic nature writing, and these essays demonstrate similar ties. Robert Michael Pyle, writing from his home near the delta of the Columbia River, discovers hidden biodiversity in a canyon full of old-growth trees, and he exults in the collaboration that ultimately helped preserve it. Sharman Apt Russell, sensuously surveying the White Sands Missile Range in New Mexico, hails that massive military base as a "biological ark," however incidental and unintentional, where mining and grazing have been forbidden. In these essays reverence for bison, condors, salmon, and bears leavens the heavy loss of *genius loci*, the legendary spirit of place that is being destroyed or driven out. It is easy for some people to make light of so-called granolas and tree huggers. But it is far less easy—considering a recent and well-known episode in the American culture wars—to ignore the passion that motivated Julia Butterfly Hill to spend more than two years in a redwood tree in an effort to rescue it. Likewise it is difficult to dismiss the passionate personal stories these writers share.

It can be a perilous endeavor, working on behalf of the environment in the West, as David Helvarg detailed in his book *The War Against the Greens*. One has to brave the power of vested interests: people who have come to count on private profits from a flow of public subsidies. Such interests, in Wallace Stegner's words, "Take for granted federal assistance, but damn federal control. They say, 'Get out, and give us more money.'" Challenging traditional rural stakeholders is tantamount to treason in small communities. Custom has hardened some land-use habits into perceived entitlements and rights. One website, maintained by inflammatory and anonymous interests, slanders conservationists as "fascists" bent on implementing "the final solution" of "rural cleansing." Rural conservationists and civil servants have been torched in effigy, downsized or fired, intimidated, even assaulted. "More than a few greens have had their houses burned down; others have lost their lives," writes our contributor Robert Schnelle. "If middle-class environmentalists are cautious, perhaps it comes from having kids who attend school with the kids of Republican ranchers." Weighing the risks and rewards of his citizen activism, Schnelle states, "For the sake of my fourth-grade son I leave monkey wrenching to others."

Most rural residents enjoy an intimate relationship with the land, but one can love a place sincerely and still be harmful to it. Public lands belong to all Americans, yet vigilante justice still takes place with some frequency today. In this new millennium, the self-styled Bucket Brigade of businessmen and ranchers protested, vandalized, and lobbied for the repeal of the Endangered Species Act in Klamath Falls, Oregon. Those renegades in the Bucket Brigade shot federally protected hawks in a fruitless move to bolster flagging pheasant populations. Federal offices there installed bulletproof glass. A year before that, a Shovel Brigade forced open a federal road in Nevada that had been decommissioned to protect endangered bull trout. Such conflicts spark righteous alliances. Militias, off-road vehicle users, tax protestors, and property-rights advocates rally with sagebrush rebels to generate a politically ticklish mix. Twenty years ago President Ronald Reagan's Secretary of the Interior, James Watt, advocated selling public lands in the western states to the highest bidder, a plan that would have led to their development and loss as public space. But resource wars and civil disobedience in defense of forests, parks, and waterways are not confined to the United States. Writing about Bowen Island in British Columbia, contributor Rebecca Raglon remembers "the adrenaline rush of struggles, when protestors sat in front of an approaching Caterpillar."

These stories reflect a global trend toward citizen activism. In response to the growing concentration of wealth and power in the hands of relatively few individuals and corporations, citizens from around the world are speaking out and organizing in their communities for a democratic and sustainable civil society. On behalf of the American commons, the authors of the present essays write of the call to activism with a passion that is based in place. They inspire engagement by means of experiences and tactics that point the way for others. These essays show that rhapsody, attachment, reverence, and awe remain vital adjuncts to the grunt work of organizing, canvassing, writing appeals, and addressing the press. As Dan Brister of the Buffalo Field Campaign confirms, an empowering outrage and a reverent hush can rise at times from the same wild spring.

Public lands, place-based affiliations, activism, and personal reflections —quite deliberately we brought together political essays of environmental encounters that cry out to be acknowledged and explored. Mitchell Thomashow, in *Ecological Identity: Becoming a Reflective Environmentalist*, confirmed the need to explore such experiences. "It is the personal introspection," he wrote, "that drives one's commitment to environmentalism." Thomashow's study suggests that experiences with nature can enhance human health; affiliations with wild spaces can help keep humans whole. As growing pains beset the West more and more each day, as natural-resource losses accelerate displacement and mobility, some citizens are suffering a malaise that has little to do with drinking toxic water or breathing tainted air. The stories related in *Holding Common Ground* need to be heard and heeded by teachers, planners, designers, managers, and mental-health professionals. Humans require contact with nature to preserve and restore their well-being. The health of society hangs in the balance. Sharing these stories is a gesture toward healing both the land and our communities.

Holding Common Ground proves that environmentalists can be idealistic enough to give up lots of time and money for their cause. Fittingly, any royalties from this book will go to two advocacy organizations in the northern Rockies: The Lands Council of Spokane, Washington, and Friends of the Clearwater of Moscow, Idaho. We editors have racked up some activist miles, Derrick for the Northwest Ecosystem Alliance and the Save Our Wild Salmon Coalition, I for the Sierra Club and a student environmental group that I advise. In the stories told here, our contributors demonstrate some of the varieties of environmental experiences in the American West today. These writers reveal, in the most intimate of terms, how they heard the call to preserve our public lands.

Section I

Awakenings

White Sands Missile Range

Sharman Apt Russell

On July 16, 1945, twenty miles southwest of ground zero, at 5:30 A.M., rancher Frank Martin was lifted from his bed and thrown to the floor. "You don't have to be very smart," he commented, "to know *that* was something new." A scientist watching the blast recalls, "There was nothing in the black desert, then all of a sudden the sky and the mountains jump out at you. Then the heat and the shock wave hit you. My feeling was simply, 'We didn't goof.'" At Trinity Site, fire melted sand to create an unnatural green and glassy floor. A mushroom cloud rose fifteen thousand feet into the air. Tiny bits of plutonium spread for hundreds of miles. The hair on cows grazing in pastures of grama grass turned white. Later, some of these cows would be sold to a meat-packing company in Albuquerque.

Scientists were not sure, exactly, what would happen when they detonated the first atomic bomb in the middle of the "black desert." They did not think, really, that the bomb would ignite the atmosphere and destroy the world. They were less certain about wind-borne radiation. In nearby towns, soldiers waited to evacuate the populations if readings got too hot. In fact, radiation levels reached 90 percent of what the army thought acceptable at that time.

For many years no one has been sure, exactly, what nuclear weapons mean in terms of our future. A generation grew up believing in nuclear war, nuclear winter, self-extinction. Fifty years later, we are more complacent. It didn't happen, after all. It hasn't happened yet.

White Sands Missile Range, home to the Trinity Site, is a three-hour drive from where I live. In central New Mexico, the range includes over two million acres of a broad basin rimmed by mountains. Here, alpine meadows and piñon-juniper forests give way to grasslands, alkali flats, salt creeks, and scrub desert. Drifting dunes of sand, pure white gypsum, flow into an adjoining national monument.

White Sands is one of the largest missile ranges in the world and one of the best. Tests, explosions, experimental flights, and top-secret projects are business as usual. Photography is strictly limited. Security, as much as possible on two million acres, is strictly enforced. Except for some game hunting, the range is closed to the ordinary citizen, to you and to me, to possible spies and saboteurs.

Ever since 1945, when the army took over the land from ranchers, the range has also been closed to cows. This is the awful truth: a well-run missile range can do less ecological damage than a poorly run ranch. Today, land inside the missile range shows less human disturbance than land outside the missile range. The best-preserved Chihuahuan Desert anywhere might be here. Wildlife is abundant. Patches of grassland look pristine.

Protection of native communities from livestock is not easy, and it is not business as usual. Fifty percent of western range-land, public and private, is severely degraded. Seventy percent of the West is grazed, including most of our state and federal lands, our national forests, our wilderness areas, our national parks, and our national monuments.

But not our missile ranges.

In the last twenty-odd years, environmentalists have discovered the Department of Defense, which controls twenty-five million acres in the United States where over one hundred threatened or endangered species still survive. At the same time, the military has discovered the Endangered Species Act and other environmental legislation.

White Sands Missile Range is now seen by some as a "biological ark," a place where rare species can be saved from extinction. The idea of such arks becomes increasingly important when we realize that over 50 percent of the Earth's species could disappear in the next hundred years. At White Sands, the Department of Defense collaborates with the Nature Conservancy, a non-profit organization, in the study and protection of the area. Around the country, in forty-one states, the Nature Conservancy and state conservation programs work on more than two hundred projects at 170 army and navy bases.

These days, we look at the big picture, not just individual species, but habitat, the places where species live. The White Sands Missile Range is habitat to a number of species listed as threatened or endangered by the state or federal government. The biologists on this range, working to save these species, have special concerns. "Hot firings" or live missiles are commonplace. Cluster bomb drops are a unique ecological hazard. Most recently, experiments with chemical weapons use a substance that settles over the desert. The chemical is non-toxic to humans, but no one knows what it does in the lungs of an animal or cells of a plant.

The biologists file their opinions. The military file their Environmental Assessment reports. Amazingly, to our credit, there is a web of laws in place to protect the web of life at White Sands. We are trying to hold on to as much life as possible. That we are doing this on a bombing range, where we test weapons meant to cause as much destruction as possible, is one of those ironies that seem to define who and where we are at the beginning of the twenty-first century.

◆ ◆ ◆

I am looking forward to my safety and unexploded-ordnance hazard training. I know of a biologist in Nevada who had to field-strip a rifle before she could study flowers at a certain army base. I know of scientists who do their work under the rat-a-tat of artillery fire. I think the mixing of cultures is amusing.

My husband works for the Nature Conservancy and is joining me on this tour of the White Sands Missile Range. We will be guided by two Nature Conservancy ecologists, both young women in their twenties. First we are sent to a room and told to watch a video. This is our training.

For fifty years, people have been dropping explosives on White Sands, picking up most of them, and leaving just a few behind. Many of the bombs do not look like bombs. Some look like toy balls or tiny boxes or darts or bullets or film canisters. Many of these bomblets, detonators, flares, and shells are small, an inch or two long. They can be found lying in the dirt or buried in the ground. Last year, on his first visit here, a nineteen-year-old airman picked up and threw a funny-looking can with a parachute attached. He died.

After the video, we fuss around getting security badges, getting a "vehicle"—as all military ground transportation is called, no matter the type—and getting our lunches unpacked and repacked. Then we are on a paved road, past a security gate, past the guard, on a dirt road, into a secret world. We have been warned. Do not photograph the horizon. Do not photograph any landmarks. Do not photograph any man-made objects.

In early fall, with the summer rains, flowers are everywhere. I see the yellow of golden-eye, snakeweed, marigold, and mustard. There are small white daisies, purple asters, blue morning glory, and pink globemallow. Slowly we bump past prickly pear, barrel cactus, mesquite, cholla, and creosote. These plants jostle each other, pushy and proud, everyone shouting. I also feel excited. Deserts in bloom have a special glory. They knock me over the head.

If I look closely, I can see the debris.

That twisted lump of wood is a target hit. That bristly silver thing is something I should not touch. Telephone poles bisect a hill. But there aren't any telephone lines here, and these are targets too: can you blow up something small between two big sticks? I see fake hills and bunkers and towers and warning signs, Danger! Danger! Danger! This is the military's vernacular landscape.

Conceptual art.

Animals are everywhere. Herds of oryx or gemsbok gallop in the distance, kicking up dust. Twenty-five years ago, these ungulates were introduced into the New Mexican desert as a big-game species. Oryx have long sharp horns and faces marked clownishly in black and white. They can weigh up to 450 pounds and they eat everything and hardly ever need to drink. They love it here. Suddenly they appear by the road, and we stop the car to stare, and they stare back, and I think: these guys are goofballs.

Then they realize we are not a tree and they wheel and gallop away, kicking up dust. I never tire of the game. I want to stop at every oryx-sighting. But there are too many of them. There may be three thousand, overrunning the range.

We see antelope too in the black-grama grasslands to the north and a big mule deer, a buck that somberly stands his ground. Golden eagles prowl the sky. We point to prairie falcons and red-tailed hawks. We listen to wrens and warblers. High in the mountains, not so easily seen, are the bighorn sheep and mountain lion.

At some point we stop and hike up an arroyo in search of the cereus cactus, which the two ecologists say is somewhere, just up ahead. The night-blooming cereus cactus is on the federal list of threatened species, endangered by collectors, grazing, and the loss of pollinators. Its flower is legendary: a sweet-smelling, five-petal star that fills the palm of your hand and seems to glow in the darkness. We chat idly as we walk on the packed sand, the arroyo winding, the flowers shouting. Morning glory! Brittle-bush! Purple aster! We walk and finally turn around, still idly chatting. Apparently, none of us ever expected to find a cereus cactus. We are all experts at this kind of thing. It's just up ahead. Oh, okay, let's go.

I am happy simply imagining this plant.

Back at the vehicle, the largest Texas horned lizard in the world watches us leave. The lizard is spiky and scaled, more alien than any Roswell story, patterned in ovals like staring eyes: lemon-yellow, orange, brown.

We drive for hours. This is very big habitat.

Along Salt Creek's edge, pupfish are everywhere. The hyperactive males, about an inch long, flaunt a pretty streak of blue. One of the rarest species at White Sands, these minnows are descendents of fish who lived in New Mexico when New Mexico was an inland sea. They exist in only four places in the world, three inside the missile range and one on its border.

At this place, in the middle of the Chihuahuan Desert, a large marshy area is fed by an underground spring. Ducks drift on a sudden pond. The oasis is green with seep willow, cattail, and salt cedar. Oasis is a Greek word, of Hamitic origin, similar to the Coptic *wahe*, the ancient book of wisdom. An oasis is a miracle. On the planet Mars, the small round dark spots intersecting the canals are called oases.

My husband is enchanted. He falls hard for certain landscapes.

The two ecologists with us also have the look of people in love. They point to the clear water roiling out of rock, to the sedge grass, to the emerald-headed mallard. They watch us carefully. Do we see it too? Do we feel it too?

All day long, everywhere we go on White Sands Missile Range, these women will show a deep, spontaneous pleasure in the natural world. Smell this leaf. Listen to that bird. Fall in love. This is a working day for these women. This is their job.

People question the importance of saving the White Sands pupfish, a species that can only live here. Many of the animals and plants that will go extinct in the next fifty years are like the pupfish, obscure, specialized, finicky.

Some people say we protect species because we never know when they might come in handy. A quarter of prescription drugs have some component from plants, and we have studied, for medical purposes, less than 1 percent of the plant species in the world. The next flower that goes extinct may hold the cure for heart disease or cancer. It may be the miracle relief for insomnia. The world is a commodity not fully exploited.

We have been driving for hours in a government vehicle. I am tired. I have a headache. I look down at the pupfish, and I am not convinced. Put him in a blender, extract his essence, the pupfish will probably turn out to be a dud.

His habitat, too, is not that impressive: this salt-encrusted ground tufted with sedge grass, this green hillock, this single oasis. The view is a little too flat for me, the sky too hazy.

A dark voice whispers. I let the words form: *I don't really care about the White Sands pupfish.*

I move my cold gaze across the face of the world.

I do not care, either, about that duck.

I look to the sky. I do not care about that golden eagle.

I scan the horizon, stranger to stranger. I do not care about that rabbit or that deer or that mountain lion. I do not care about these hills or these undulations of native grama grass. I do not care about the finger-lock meadow, the ancient creosote, the blue-green lichen.

Once we choose not to care, how do we ever stop?

We can start naming the list of things that no longer exist. We can live in some reverse mythology, not a sense of the divine naming Earth's creatures, but a sense of ourselves in a ritual of unnaming, coldly watching, undoing the world, and not grieving.

We can care or not care about the extinction of a species and perhaps all we will ever know is what that means to us right now, whether caring makes us feel more alive and not caring makes us feel more dead.

Suddenly frightened, I admire the pupfish. I look for the beauty in the salt creek, in the sedge grass, in the milk-vetch. I am trying now to save myself. Yes, I agree. It's wonderful.

We get in the car and drive home.

Farther north and east is the Trinity Site, which opens to visitors twice a year. On those two days, once in the spring and once in the fall, thousands of people from all over the world enter White Sands Missile Range and drive past the guards. At a parking lot in the middle of nowhere, they leave their cars and buses. A sturdy cyclone fence surrounds the site, which is not large, the size of a football field. Visitors are warned not to pick up the tiny pieces of green glass that can still be found littering the ground like radioactive confetti. Ground zero is marked with a plaque. A display tells the history of the atomic bomb. Outside the fence are booths selling food and souvenirs. There is hardly anything else to see but the rolling desert.

Thousands of people come every year.

You don't have to be very smart to know this is something new.

We didn't goof. That's debatable.

In any case, we stand now in the middle of a desert that was once an inland sea. The pupfish and the cereus cactus are somewhere nearby. We may feel a little confused, here at the beginning of the twenty-first century. Maybe we are wondering what to do next.

Or maybe we have already fallen in love, and it's just that simple.

The Middle Ground

Sherry Simpson

I lay belly down on a platter of rock and dipped my hand into the stream's surging waters. I was thinking shades of blue. Aquamarine. Turquoise. Azure? But there is no word for distilled sky. The best I could do was imagine the tincture of Aqua Velva aftershave, the exact odd color of Sheep Creek.

A shout from my friends roused me. We had just begun our hike into the valley, and already I was distracted. Kathy and Kim each grabbed a hand to haul me up the slippery bank. We were not exploring this mountain cleft for pleasure, exactly. Juneau's coastline is embedded by other valleys. But this place was different. We had come to look carefully, to appraise loss, to affix meanings. We thought it would be the last time we'd see the valley. It hardly matters now that we were wrong.

On that rare clear day, Sheep Creek Valley unfolded before us like a magician's trick. Snow still thatched alpine slopes, and ridges crimped the sky. Five miles north lay Alaska's third-largest city, Juneau, but you would not guess that while walking on the toes of Mount Hawthorne and Mount Roberts.

A dark-haired man wearing a red bandanna around his throat hailed us from the stream. He was helping a boy with a fishing pole. Ten years before he had hiked to this spot while hoisting a fifty-pound pack, he said. "Nothing here but mine ruins then," he added, lofting his voice over the stream. "Now there's a mining outfit up there."

As if it were news to us, he said a gold miner from Echo Bay Alaska, Inc. had told him that protests wouldn't matter—the dam would go in for sure. The man jerked his thumb at the boy, who was about seven, the same age I was when my family moved to Juneau. "I figure this will be a momentous occasion for him, something to remember when he's old," he said. The boy, intent on rigging his pole, never turned around. "We're going to catch the last trout!" the man shouted as he waved us on.

The season had stalled just after high summer. False hellebore slumped by the trail in brown, gloopy streaks. Chest-high ferns feathered in the breeze, and fireweed extinguished itself in cotton batting. It is hard enough to memorize something as blunt as a mountain, as contained as a stream, much less all this fine, rampant detail. But I looked closely, and I wrote things in my notebook—"drowsy smell of vegetation," and "elderberry clusters," and "devil's club"—as if words were ever sufficient.

Words were all I had. I covered mining for my hometown newspaper, thrashing my way through jargon-choked thickets of environmental-impact statements, consultants' studies, agency reviews. I attended public meetings where nobody ever convinced anyone of anything because everyone was preaching to the choir and calling out for amens. I listened to environmentalists and took notes, and then I called up mine officials and took more notes, and finally I typed out absurd observations for the public, such as, "Figuring out what the mine would mean to water quality, fish, wildlife, and marine organisms won't be easy." What else could I say? I was only the translator, and not a particularly good one.

I told people I had not decided whether I favored or opposed the mine. That was my job, and it seemed true. Juneau's history is a history of mining, after all. Sheep Creek Valley was named by the town's founding prospectors, who recognized gold but not mountain goats. For fifty years, three of the world's largest gold mines operated in Juneau's back yard, along with many smaller outfits. The dam was part of a plan to reopen the old A-J, a project that promised more jobs and less reliance on oil, government, and tourism. Nothing new about this story.

We ambled through groves of old black cottonwood trees, stately and sweet-smelling. Southeast Alaska lies almost completely under the dominion of spruce and hemlock trees, and almost nowhere else around Juneau do cottonwoods thrive as renegades from the rainforest. Environmentalists argued that the trees made Sheep Creek Valley "unique" and hence worth saving. It was a small irony that these exotic trees probably would not exist if not for the first miners who cleared the valley floor in the late 1800s.

I found myself totaling up signs of inhabitation and wilderness. Here, a rusting shack carpeted by porcupine droppings. There, a giant platform bulldozed from waste ore, engulfing cottonwoods that now rose from stone. I counted leaning telegraph poles, and noted obscure bits of corroded metal, and listened absently to mechanical rumbles echoing across the valley. Close by, a varied thrush struck the single bell hidden in its throat.

Nobody saw the same things in Sheep Creek Valley; nobody knew how to speak about what they saw. One environmentalist expressed typical sentiments in a letter to the editor urging the company to dump the tailings in "another much less valuable valley." She meant a valley farther from town, one we wouldn't have to look at, and one sure to be wilder. Pro-mine factions used their own rhetorical topology: The valley's history meant it was no longer "true" wilderness, so damming it would be no great loss in a landscape reamed with thousands of "pristine" valleys. A mine consultant said: "There is a whole lot of emotion swirling around this project. It would be nice if it was based solely on science," as if science is the language that would purify the discourse, making everything simpler for everybody. I could see why he preferred numbers, though; words like "mitigation" and "monitoring" and "reclamation" sprang from a deviant vocabulary far removed from the green actuality of this basin.

All through the valley, life flickered at the edges. We scanned the brushy slopes, looking for black bears. Crossing a quiet slough, we startled a Dolly Varden, a shimmering, sea-going trout named for a Charles Dickens character. The perfect, tiny track of a Sitka black-tailed deer inscribed moist earth. Where Sheep Creek spreads itself into shallow riffles, we paused to watch two plump American ouzels flitting along the bank. One of the slate-gray birds marched into the stream until completely submerged, and I longed to see it striding underwater against the current, a bird comfortable in two mutually opposing worlds.

Just before the trail scrabbled up the mountain, we paused by the creek. Kim dipped bare feet into an eddy and gasped at the chill. Kathy glassed a cliff for mountain goats. I contemplated a fire ring littered with empty snack cans of Libby's Peaches Light, trying to calculate the gap between discarding an aluminum can and discarding a mountain of tailings.

I had been afraid to care about this place too much, the way you might hesitate making friends with a terminally ill person. You know it would be meaningful, that perhaps you could do some good and even learn something, but you also know that in the end, you will reap pain and loss. It was too late, of course, but, looking at those Libby's cans, I realized I was against the mine and had been all along. The journalist's creed of objectivity made a convenient shield fashioned from empty words carefully chosen for neutrality. It had become so easy to hang back, to let others squabble on the opinion page and in public hearings. I had made a career of weighing pros and cons and forever fiddling with the scale, trying to make it all balance evenly. The middle ground—that was my territory.

We walked back faster than we came. The sun had dipped behind Mount Roberts, and blue shadows avalanched behind us. The stream seemed unfamiliar from this direction, and so I was startled when we again encountered the fisherman, now setting up a red dome tent beside the creek. He looked up as we passed and shouted, "We caught eight trout!"

We climbed the final ridge that separated us from the rest of the world. The throaty percolation of Sheep Creek fell away behind us. As we rose above the valley, I thought, "I should look one more time," but I could not. We crested the ridge and descended into the darkening forest without turning back.

You could say this is a story with a happy ending. Not because I left newspapering so I could say what I wanted to say (I did), not because the whole town rose up against the mine to save the valley (it didn't), but because the price of gold dropped and the company pulled out. This was a temporary triumph, if triumph is the right word. Today townspeople argue about other problems—cruise ships discharging wastewater and smoke, and eco-tours clogging the skies with flightseers and the backcountry with people. Mine gold, or mine scenery—the results aren't so different.

And there is, of course, what we bring upon ourselves, without protest, hardly without noticing. Juneau's forests, valleys, and wetlands disappear a scrap at a time, paved under subdivisions and malls. Sheep Creek runs clear, but few other streams survive in their original state; they run through concrete culverts, or dribble through trailer courts, or drain sluggishly through silt and garbage. It must have happened this way in other places;

one day you thought there was enough land, or wilderness, or whatever you call it. And on another day, you realized there isn't.

Sometimes I imagine lying next to that stream of unnamable blue, face pressed against the damp earth, listening to the low song of the stream humming against the granite floor. Days and nights would rain down, burying me deep in the abyss of history, thinning me into a sedimentary layer that compresses place and possibility. The sun slanting over the mountain's brow, the warm stirring air, the endless prayer wheel of the stream's passage, and life greening all around—this is beyond measure, beyond words.

Sheep Creek taught me that knowing your own heart is not enough to save the world. I confess I don't know what is. But if words are all you have, then what else is there to do?

Northwest Passages

Laird Christensen

It is hard to imagine, looking up the Columbia River gorge, a power that could hack so wide a path through sixty miles of mountain. Auburn walls of grooved basalt rise straight up from the river, and streams that drain volcanoes drop hundreds of suddenly vertical feet. White strands dangle, shimmering from the sharp edge of forest.

The gorge recalls a Northwest locked in ice, when rivers from the northern Rockies backed up to form an inland sea called Lake Missoula. Before the world began a steady warming some twelve thousand years ago, ice dams formed and buckled dozens of times, and silver cliffs of water shouldered through the cracks, shattering down the Columbia's channel. The floods ran a thousand feet high as they blasted through the Cascade Mountains, monstrous waves recoiling as ridgelines broke and toppled.

Today, Interstate 84 trails a quieter Columbia east from Portland through the gorge, then slides across high desert toward Boardman before leaving the river for Boise and the Great Salt Lake. As it passes only a mile from my family's home on an ancient floodplain, it's a road I know well. It led to my family's camp up the Klickitat River, and it was my escape into rain shadow during months of winter gray. Over time it came to mean freedom, for it was the quickest route to a world beyond Portland's swollen appetite and toward an identity less burdened by other's expectations. On the day I finished high school, it led me away from Oregon.

Even as a great-grandson of the region's white settlers, I never really knew the Northwest while growing up. Public education, popular culture, and white-bread religion all helped shape a childhood that could just as well have unfolded beside the Chesapeake as beneath the Cascades. Ours was a sawdust-sweet house in one of those neighborhoods that appear where the restless fingertip of suburbia jabs between farmland and scraps of forest. Ponies grazed past the end of our fresh-paved street, and in the woods behind their pasture were deer, fox, and more blackberries than I could ever hope to eat. But even my experience of wild places was homogenized: those dripping western forests were merely a stage—convenient, but arbitrary—on which to act out frontier fantasies. By the time I left home, what woods remained in the neighborhood were giving way to strip malls and subdivisions.

I spent the next nine years drifting over the continent, taking any work available and trying on identities against a hundred passing backdrops. If I stayed in one place more than a few months, the urge to be back on the road grew insistent—it was being between places that seemed to me most satisfying. Gradually, though, the thrill of arriving so often in a new place gave way to a different impatience. At twenty-seven, I paused for college in rural New Hampshire, staying long enough for roots to take hold in the leafy hills near Mount Monadnock. In the years that followed I began to understand at last how one becomes a part of place.

Gradually I saw that place is no mere setting, but a fabric woven of many lives and processes. Learning my role in the Monadnock region began with the acknowledgment of my neighbors. At first it was enough to know their names—to tell striped from sugar maple, fisher cat from weasel—but seeing your neighbors as individuals won't get you very far in a world made up of relations. Working summers as a ranger, I learned that wild rhododendron flourish in soil made acidic by the decay of red oak leaves and hemlock needles. They do not simply depend on these neighbors for their health—their very existence is a community relationship.

Understanding other species as expressions of community eventually led me to see myself in the same way: I am not some rootless individual who takes in water and protein and minerals at his pleasure; rather, I *am* locally specific cycles of water and protein and minerals. The more I learned of plant associations and wild populations that overlapped to define the watershed, the better I understood the role of my species in this place. The better I understood myself.

It was while discovering the Monadnock bioregion that I realized how little I knew about my native land. I had no doubt that the Northwest had shaped me, but I had no idea how. So, when the University of Oregon offered me a fellowship in 1994, I was ready to get to know my native community for the first time.

❖ ❖ ❖

In the coming years I spent all the time I could in those old Cascadian forests. Before long I knew enough of my neighbors to begin to feel at home among them, but from the very first I found that I already understood this place better than I could ever know New England. Particular combinations of colors and textures evoked a startling, bone-deep intimacy: the damp glow of mosses against the russet crumples of basalt; the sheen of slow water stained that particular green of cedar shade. Each time I stumbled into such a perfectly familiar mosaic, I felt unaccountably safe. I belonged here.

But intimacy has its consequences—particularly in a land under siege.

One crisp October afternoon in 1995 found me tracing the Willamette's middle fork past the logging town of Oakridge, and then up Salt Creek into the Cascades. At forty-five hundred feet, the autumn sun was plenty strong, highlighting the range of greens along Bunchgrass Ridge: western hemlock, red cedar, and the ubiquitous Douglas fir. Beneath the mesh of conifer boughs swirled ferns and sprawls of Oregon grape.

Resting above the Warner Creek drainage, I looked east from the shoulder of Forest Service Road 2408 to where nine snowcapped volcanoes rose like vertebrae from the spine of the Cascades. They shone so bright against the plush sky that the scene would have fit well on a calendar—if not for the carnage in the foreground. Without turning my head, I counted forty-two clear-cuts, oddly geometric gouges torn from the flanks of forest.

Like most roads in the Willamette National Forest, 2408 is a logging road, for the Forest Service is in the business of selling public timber—at a loss—to corporations. Over the years scientists and citizens have begged federal protection for what little remains of our temperate rain forest, and the Warner Creek watershed had been set aside as habitat conservation area for the northern spotted owl. However, as a result of arson and a salvage timber rider tacked onto an unrelated spending bill, it was due to be logged come spring.

After eight miles, 2408's resemblance to the thousands of logging roads that jigsaw public lands stopped suddenly at the makeshift border of *Cascadia Free State*. A wall of vertical logs, ten feet high, spanned the road beyond a dry moat. Across the drawbridge, past the palisade and the tepees that housed forest defenders, was an inspired array of roadblocks, ranging from rock walls to ten-foot-deep trenches. Forest Service gateposts were covered by lockdown barrels—fifty-five-gallon drums filled with concrete; defenders could reach through horizontal PVC sleeves and lock the carabineers on their wrists to the hidden posts.

Work on the road continued day and night. While some took pick and shovel to the trenches, others took pry bars to the ledges above to dislodge oven-sized boulders, which tumbled through the dust of tiny landslides and thudded onto the road. A half-dozen waiting men and women would gather then, inching the boulders toward walls that appeared overnight. An exuberant defiance filled the encampment, but I also felt the darker strains of despair, fury, and hard-earned paranoia.

Although I had discovered what it meant to feel at home in Cascadian forests, so much of my experience of Oregon has been stained by conflict between people who love the land in such different ways. Those who work in the timber industry often live much closer to wild land—and know it better—than activists who come fleeing urban clutter. But many people in logging communities like Oakridge do not look past their paychecks to the web of external factors—foreign trade policies, corporate colonialism, and other political power games—that controls their lives. At the same time, too many environmentalists forget that shifts in forest policy, no matter how justified, force people out of patterns of behavior that are familiar and dependable. Antagonism thrives in the space between their experiences, haunting this clear-cut patchwork.

The perspective that I brought home to Cascadia, that humans are but one part of a community of life processes, shapes the way I see these conflicts. It is painfully clear that current logging practices devastate these forests, and so the safety that I feel here is a paradox, for my community is disappearing. At each clear-cut slope—and I have seen more than I care to recall—I am reminded of my kind's short-sightedness. Grief and anger tangle deep in my throat. Sometimes I miss the third-growth forests of New England, which have spent more than a century covering the scars of human abuse. Each stone wall threading the smooth pillars of a beech grove is the frayed collar on a once-tamed dog, which I love to watch run wild.

But I cannot turn away from this destruction of my native forests. Finding yourself at home means accepting the obligations of community, as well as enjoying the privileges. If we see ourselves as plain members and citizens of a land community, as Aldo Leopold did, then some of those obligations become clear. We have a responsibility to speak for those members of the community whose voices are not heard. We have a responsibility to influence public policy to reflect the best interests of the entire community. And, like the activists who defended Warner Creek, we may also have a responsibility to defy the letter of the law when misguided policies threaten the health of the community.

If we focus on protecting the forests, however, without also caring for the humans who make their living there, we have settled for too narrow a definition of community. Many loggers have only the best interests of their families at heart, and there is no disputing such motives. But there are ways in which timber workers can continue to support their families while also allowing the forest community to recover. After all, we will always need lumber, and we have the power—as consumers and as participants in democracy—to demand a forestry that is based on respect for the entire community.

The day I scraped the *Stop Clear-Cutting* sticker from the bumper of my truck was the day I realized that sharing the vision of sustainable forestry demands open and honest communication, and that grows best on common ground. While the need remains for triage activism, we will best protect wild lands by helping others understand that our own health, the health of our families, and even the health of our economies depend ultimately on the health of the larger community to which we all belong.

Then we will truly have found our way home.

Going Wild

David Axelrod

Calendar photographers and tourists do not gather in Oregon's Elkhorn Mountains. The crowds favor the Wallowas, east across the valley, a more conventionally picturesque wilderness. Not as high, the Elkhorns have less moisture, less geological complexity, less topographical diversity, and less protection from formal wilderness designation. Mining, though meager, continues decades after the Oregon Gold Rush; crossed pickaxes dot maps. Mine shafts, tailings, and ponds polluted with heavy metals are still found on the ground.

Having lived my early life far from the Elkhorns makes me, in the eyes of the descendants of early settlers, an interloper. To live at ease with the "traditional values" of northeastern Oregon means accepting as right and proper the violence that wounded these mountains. These are public lands, but "public" in name only, and the "right" to make a living off of this land is retained by those who arrived here in the second half of the nineteenth century. Nevertheless, my early experiences in the Ohio coal fields, where uplands and valleys vanished under tailings, causes me to feel oddly at home here, to see clearly enough, despite my disreputable status.

The Ohio, with which I was intimate—an area running diagonally across the eastern third of the state from Nelson's Ledges to the Hanging Rock Iron Region—provided ore, timber to fire furnaces to smelt the ore, and later coal to fuel heavy industry iron made possible. That former wilderness exhausted its resources and today remains impoverished. No longer valued, the land ignored, wounds inflicted on it have begun the impossibly slow process of healing. I came of age walking at the edges of those wounds. In the Elkhorn Mountains, as in other areas in the interior Northwest, I walk at the margin of a peculiar violence. I have seen it before; its consequences are familiar.

From Dutch Flat Saddle in the Elkhorns, it is a short descent to Cunningham Saddle, where the trail crosses the ridge and looks down on the headwaters of the north fork of the John Day River. The upper reach of the watershed is a timbered valley—lawfully designated wilderness. Three miles farther downstream, however, logging and mining define its western boundary. Beyond those cut-over areas, parallel mounds of tailings line the stream until it disappears into a canyon, entering yet another portion of official (though fragmented) wilderness. Along ridges that surround the drainage are familiar patchworks of clear-cuts.

Can such fragmented lands in any legitimate way be called "wilderness?" If not, landscape is reduced to an esthetic perception, that is, to wilderness defined only by one's ability to look across a significant distance without any evidence of human intrusion—an illusion the western landscape allows in some places. Such a definition, however, condemns these fragmented lands to further neglect, even contempt.

Or we may concentrate on the near at hand, which is wild, "innocent" of such aggressive desires as logging or mining. Here in the Elkhorns, the intimacy of looking at the wild near at hand, the specific living being that refuses to abandon its place, becomes an awkward necessity, the sole criterion for defining wilderness. Still, whenever I look up from the ground at my feet, contrasting shades of light demarcate the bounded wilderness and the landscape transformed by industrial uses that surround it. In some places, the wilderness boundaries are clear-cut surveyor-straight.

There are always more edges to the inviolate wild, ever more bounded territory where evidence of human intrusion cannot be forgotten, much less ignored. But to long for a former world or pretend it still exists is a delusion. Ours is a world diminished by our aggressive inventions, and to deny that is to willfully participate in the ongoing crime against what intact landscapes remain. Our delusion is predicated on the notion of a receding frontier and a historical territory beyond that frontier; with enough dollars you may visit before it too is "corrupted" by modernity.

The lonely planet, with its five billion hungry human souls, is not lonely, and "unexplored territory" is nothing less than a ploy meant to appeal to the Romantic longings of the privileged. Few of us live in a place that remains anything like its remote, inviolate creation. Our world is more ambivalent and complicated, a place of broken light and shadow observed only in contrasting, dissonant elements. Where I live is a landscape that, though it is not urban, is nevertheless wounded by industry, the flow of profits from which leads directly back to the cities that many contemporary rural (and urban) dwellers say they disdain, as though disdain of this sort were a crucible in which we are made pure. Does virtue accrue to us because we are willfully blind, disingenuously ashamed of the diminishing beauty of the natural world? Does such "virtue" exempt us from Earth's judgment?

My former student, Jerry, was the child of a family that came into these northeastern Oregon mountains soon after arriving here in a wagon train. They homesteaded in the Baker Valley and for three generations mined claims just south of Summit Pass. He remembers how, on a lark, they would dynamite streambeds that once were full of salmon smolts from the sea-runs that spawned in the creeks on their claim. He remembers sunlight refracting through the spray, the heavy thud of boulders landing on moss and old ponderosas crashing to his and his father's and uncle's cheers of triumph. Before they destroyed those creeks, he recalls the water turned white with milt during the spawn. He grieves now over his participation in that greedy fantasy of quick wealth that never came. The perfection of that earlier world he remembers is lost, the scars of his passing visibly evident to him from many miles distant. He never returns to these mountains, never risks any intimacy; his own life is strange to him now and strangely impoverished.

Jerry has since moved to Kansas.

My family has lived in Ohio since the eighteenth century and played an early role in its being raped, never asking whose land it was before we claimed it ours. Instead we told the story of hardship endured, the winter men quartered at Fort Defiance, waging Mad Anthony Wayne's Indian War. The women left behind at that edge of Ohio wilderness failed to gather the garden before an early freeze. Cows fell ill, then children. Game grew scarce and famine walked back roads. An infant died. Perhaps others would die. Perhaps we were observed gouging out a grave in frozen ground.

Observed by whom? Wyandotte, Miami, Shawnee, Erie? She could have been any of them. The only word we had for her then as now is squaw, an insult. She appeared, a trick of sight, from the forest across the field from our unpainted saltbox, a deer slung across her shoulders, and staggered over snow toward the door, my kin cowering below window sills, loading muskets, giddy with the imminence of battle. She lay the deer at the threshold, where I have stood many times, listening to corn thicken in maniacal rows, where I have caught only a glimpse of her absolute retreat across that field two centuries ago.

Who was she? Even if I knew—forests beaten back now into hollows, old trees gone, creeks fouled by mines and drifted deep with silt—would I dare offer her thanks? I wish to find an end to this legacy of cruelty, but wherever I walk, I walk the edge of a wound our history seldom permits to heal.

The Wrights farmed east of our homestead after the Civil War. They were brother and sister, ex-slaves who found one another and for the remainder of their lives honored the primacy of blood. Our last memory of them is threshing time circa 1900. No one knows who buried them side-by-side in a plot at the back of their farm, a pasture overlooking woods that surrounded their cabin. All that remained of the Wright's cabin was moss-covered limestone foundations grown up in vine maples and orange touch-me-nots I was taught to crush to soothe a rash.

Their forty acres became ours, though no record of sale exists. Thus we took possession of mineral rights to strip-mine that hillside, though no one bothered to scruple about the overgrown graves as lights from the big shovel shone above the ridge at night. We "reclaimed" the land afterward: leveled the hillside, trucked and bulldozed topsoil back into place with its cargo of human bones, right and proper for the sowing to soybeans and corn.

I once hurried across the sheep paddocks, climbed tailings replanted in parallel rows of white pines to a clearing overgrown by clumps of grass I knew only by the name "poverty grass." Thickets of sumac and blackberries teemed in the valley, scabbing over the flayed carcass of hills men had gorged on with their colossal machinery, hauling away coal. A kind of revelation sweated through the pores of my skin, a legacy that permitted no exceptions. The terrible gift of that land: healing, impudent, lovely. A mind was going wild.

We pretend that our relationships to the land are somehow different in the western than in the eastern United States. The West is endlessly rich with possibilities: scenic, economic, spiritual, all of these as self-aggrandizing as the myths told about any frontier. We live as Romantics. One writer fantasizes about being the first explorer of the upper Snake River Valley, the first to walk under the Tetons. A feeling arises from that image of impossible, loathsome longing for a "virgin" world, a "true" wilderness, innocent of history. No such world existed, only another people's way of living in relation to what we mistook for an uninhabited land. Wilderness is not a status we grant the landscape. It is Earth's only way of being as it chooses. What it chooses to be is alive, and if cut to the quick, to knit itself again into a whole.

Let Romantics walk in isolate dreams of a static virginal world. I wish to walk only here, where what is alive is most thoroughly and undeceptively alive: these wounded mountains.

Slate Mountain

Will Peterson

A pair of Cooper's hawks came back each spring to the old Doug fir above Slate Creek. They were quick to show their displeasure when I appeared those years on the trail below and then reappeared from the upper canyon. I couldn't blame them. Their prospects looked pretty fine: the east-facing slopes of Slate Mountain all the way to Mink Creek Road, with only the occasional interference of a golden eagle or two drifting from the Snake River Plain.

So I drove up that first of May with a mind toward seeing if they'd come back, and as a subplot, to see if I could make the grade at all. I'd ruined my hip three years past by going too fast on an icy mountain trail—had gotten up and fallen again—and the only question was how much worse it was going to get.

Slate Mountain is actually a quartzite ridge, folded and tilted perpendicular at the northern rim of the basin and range province, just another example of the arbitrary nature of naming. But at the trailhead, Mink Creek has undercut an area of evenly bedded slate; and visitors have accelerated the process by gathering shards that break and slide creekwards. Dwarf phacelia, scorched penstemon, and ancient juniper seem the only natives particular to this terrain. But thirty yards south and ten thousand years earlier in the career of Mink Creek, the more gradual slope has allowed the flourishing of wild rye and roses, small groves of black hawthorn and aspen.

A breeze was blowing through the willows of Mink Creek as I got out of the car. A goldfinch wove through their upper branches. I might have stayed there, but I turned instead and began the trail.

It headed straight up to the slate peeling off the hillside. Barely able to bring my left foot level with my right, I had to stop after twenty yards and look around. There's always a tradeoff. The scenery here was just as pretty as a mile up. I paused in a grove of young black hawthorn, the only tree able to withstand the overgrazing endemic to southeast Idaho. Its thorns are beloved by winter grouse, which roost without fear in mature stands. Then in another ten yards, I noticed for the first time a juniper clinging to the slate ridge. No more than five feet tall, it was at least two and a half at the base; it seemed to be keeping to the ridge by force of will and root alone.

Placing one foot before the other, I entered the mouth of the ravine where the slate sloughed down on both sides. Higher, I knew, was a basin where birds of passage hung out among the chokecherry, dressed up for Saturday night and singing like it. And higher than that, beyond the second switchback atop the ridge, were the Doug fir and the hawks.

Already I could hear the songbirds farther up the trail: the goldfinch, western bluebirds, the horned larks and sparrows. So much for the old saw about birds of a feather: with a pair of Cooper's hawks around, there was a lot of multiculturalism. I never knew when I would encounter the hawks in the canyon, but when I did, it was magical. As Black Elk said, "Theirs is the same religion as ours."

But pain's message, though non-verbal, maintains a certain authority. I wasn't going any farther. I stood and watched how the slate peeling off the opposite hillside flowed wave-like to the bottom of the ravine and only there achieved turbulence. Near at hand, where the trail had been cut, the same process was at work. The layers of slate lay in alternating shades of gray, gray-blue, and violet.

Only because we fly here and there like the Red Queen in *Through the Looking Glass* do we not see it. Only because our heads are chinked full of distinctions between living and non-living, between brief time and circular time, do we not believe it: the Earth is alive. Slate slides out of the locked hillside to become available for life. This grand flow of life and energy is frozen only to those who won't slow down long enough to see it. Is our only holy hope that someone will set us down hard on our butts long enough to pay attention?

I'd gone maybe a hundred yards. I started down the trail, and I couldn't go any faster down than I'd gone up. A hiker or mountain biker looking on might have felt a twinge of pity. Although it's one of my favorite emotions,

I didn't feel self-pity. I knew the Cooper's hawks were up there, on the other side of the ridge, watching Slate Creek. I didn't have to see them.

Sometimes you can't go to the places you love.

From the time I was a three-year-old runaway, all I needed was a peanut butter sandwich in my pack and someplace I'd never been. How many trail sides passed in a blur? I'd always linked freedom to physical mobility—why not rhyme virility? Is virility a measure of how far you can ride, hike, or run through a landscape, or the resolve to ensure its biological integrity? I don't see a false alternative here, but the question must be considered, as the idealists of the 1960s quickly age.

Sometimes you can't visit those places you've heard about, where white bears roam in a green forest or sacred nations of caribou journey east of the sun and west of the moon. But you've done something for them. You know they are there.

I got back to the car. The wind was blowing from the highlands. I took one look back. Across the brow of the cliff, between the twisted arms of the old juniper, a Cooper's hawk came kiting. She rode the wind across the mouth of the ravine and then the brow of the next hill, pitching wind's velocity to Earth's pull, made a wing adjustment here and there among the aspen and chokecherry, and then she was gone.

For Jackie Johnson Maughan

Full-Stomach Wilderness and the Suburban Esthetic

Harold Fromm

> These wild things, I admit, had little human value until mechanization assured us of a good breakfast.... When we see land as a community to which we belong, we may begin to use it with love and respect. There is no other way for land to survive the impact of mechanized man, nor for us to reap from it the esthetic harvest it is capable, under science, of contributing to culture.
>
> —Aldo Leopold, from the Foreword to *A Sand County Almanac*

I was inducted into the environmental movement in the early seventies as a result of an idiotic move to a seemingly idyllic farm located only fifteen miles south of the steel mills of Gary, Indiana. In those days I was not alone in being innocent of the fact that pollution traveled not just fifteen miles but fifteen hundred miles and more. But the resulting nightmare, illnesses both bodily and psychological, transformed my life and recruited me into the ranks of the ecologically committed. I wrote about and waged campaigns against air pollution and the general depredations of corporate environmental destruction. After an escape to the northwest suburbs of Chicago in North Barrington, Illinois, I continued my activism, this time not only with regard to industrial pollution but to pesticide spraying for mosquitoes, leaf-burning, water contamination by run-offs into wells and aquifers, and so forth. The village trustees hated me. But despite this ecological commitment, I never identified with terrorist types—animal-rights fanatics who destroyed laboratories and opened cages on family farms, ruining multiple lives in the process, or Earth First! types who spiked trees climbed by actual human beings like us, who are maimed for life. I found Dave Foreman's remark that he would sooner shoot a man than a grizzly rather far over the top (though Foreman, like other radicals from the sixties and seventies, has since morphed into a pussy cat). To me, these were self-involved narcissists, no better than the bombers of abortion clinics and murderers of physicians.

But it's probably safe to say that even extremist types have done some good in jump-starting the reforms of society. The trouble for me is that relentless one-string activists approach too closely to religious fanaticism; they are too certain of pious absolutes that tomorrow will be seen as personal pathology. Ecological Jerry Falwells are just not my thing. So when it comes to wilderness, I'm suspicious of uncompromising purities—even when they come from Thoreau or Leopold. I've said and written before that I don't believe anything human can ever be other than anthropocentric—and that biocentrism is just anthropocentrism in pious drag, like Jerry Falwell telling us what God wants, a God who always turns out to have the atavistic brain of Jerry Falwell. Of course, there are different varieties of anthropocentrism, some more benign than others.

So the sentimental extolling of wilderness found in a book like Max Oelschlaeger's *The Idea of Wilderness* never really appealed to me, and his golden-age view of hunter-gatherers seemed preposterous. Still, wilderness was a fairly abstract thing during my thirty years in Chicagoland's prairies, and it was not until my move to Tucson in 1998 that it became a concrete reality intersecting with daily life. If I was skeptical then about *wilderness = pure; society = impure*, I am now a total nonbeliever. I live on a ridge in the foothills of the Santa Catalina Mountains north of Tucson, where every day and every sunset are spectacular. I can testify to the fact that Tucson is surrounded by vast areas of wilderness, most of it quite inaccessible, despite the foot trails that afford entry into the mountainous areas closer to town. Between my home northwest of the city and the Pinal County seat of Florence, southeast of Phoenix, lies a forty-five-mile stretch of back-roads desert that is unsettled enough for me finally to have bought a cell phone. In the event of an auto breakdown, a scouting party attempting a rescue will not have to discover my bleached bones.

The Coronado National Forest spreads its discontiguous immensity all over southeast Arizona, and vast national parks and Indian reservations occupy much of the state. Between Phoenix and Flagstaff and Prescott lie additional uninhabited and majestic vastnesses. The eastern United States may be packed, dense, and built out with habitation, but once you cross the Mississippi, the wide-open spaces are not purely mythic. Much if not most of this area will never be hospitable to settlements, and the environmental mentality that bit by bit is spreading its influence offers greater and greater resistance to wildcat development, even if it's too late to save Phoenix. Tucson has been remarkably resistant to freeways and massive urbanization. Even today it has qualities of paradise, plunked as it still is in the middle of wilderness, though a realist view would prepare for its inevitable San-Diegoization, as the pace of building speeds up along the beckoning corridor to Phoenix. But what's left is immense nonetheless.

It has become a truism that the wilderness is a modern invention, an esthetic object, a rhetorical device that didn't exist for people who inhabited it when there was nothing else. When I drive around the Tucson suburbs and view the astonishing beauty of the mountains or hike in the trails of the Catalinas, I am struck by the fact that it is often suburban development that has freed up, even created, the breathtaking vistas for which Tucson is famous. The fabulous sunsets seen from Gates Pass Road and the shimmering beauty of monsoonal fogs on the Catalinas have been made visible for us by the accomplishments of pampered, technocratic, impure bourgeois like ourselves. In contrast, to walk through the ruins of the Hohokam Indians (500-1500 A.D.) in Catalina State Park is to marvel at a life that appears almost impossibly brutal amidst the dry, burning summer heat and seeming absence of shade, water, food, and websites. Could there have been substantial periods of "quality time" during which these environmentally challenged people sat around and exclaimed over the scenery like us, while shamed by their good luck at being mostly skilled farmers instead of full-time hunter-gatherers? Or, too oppressed for leisure, did they instead secretly harbor inchoate longings to become twenty-first-century bourgeois—connecting with enough "nature" at the Arizona-Sonora Desert Museum followed by high tea at the Tohono Chul Tea Room—that they could really begin to enjoy the wonders (instead of the obstacles) where they lived?

I contrast the fears of that primitive habitat with the undeniable appreciation of their surroundings displayed by the residents of my community of SaddleBrooke, nestled in the Catalina foothills about twenty-five miles from downtown Tucson in the outermost fringe of suburban development. Retired into the leisured life of affluent bourgeois, they seem to have world enough, food enough, and time enough, given our extended life spans, to appreciate the value of the "unspoiled," of the "domesticated sublime," as William Cronon aptly describes it in his definitive essay, "The Trouble with Wilderness." I marvel at the paradox: as suburban development spreads into the wilderness, it both destroys and produces it at the same time; as the wilderness recedes further and further, it becomes an object of contemplation to be valued rather than feared. Only then do mountain lions and bobcats acquire an autochthonous beauty that fills us with Prufrockian guilt about disturbing "their" universe. How beautiful could it all possibly have been to the Donner party, trekking from the Midwest to the West Coast without sports energy bars, SUVs, cell phones, or coats of Polartec fleece from Lands' End? Those who survived needed to be us to appreciate the beauty that sheer surviving made impossible for them to see.

Nowadays, some of the most remote and undeveloped wilderness can be explored and enjoyed through the prostheses afforded by contemporary technology, from motorized vehicles (a blessing as well as a curse), water purification equipment, and space-age clothing, to freeze-dried food, propane cookers or sun-ovens, and cell phones, conjoined with the reassuring sense that one has a home elsewhere, a safe haven, when supplies run out. When all else fails, helicopter rescues are yet another twenty-first-century entitlement of Everyman. Thus, the social and the technological give us more and more of a highly valued wilderness they seem to be pushing further and further away. (I refer the reader to Cronon's essay for an exemplary account of the myth of purity and pristinity that undergirds the putative sanctity of the untransgressed.)

An open-ended negotiation between development technology's creative and destructive forces may be the only resolution of what qualifies as cohabitation with "the natural." And even what constitutes creation and destruction is hardly self-evident or clear. As I emerge from the supermarket to sensational mountain vistas, or sip beer with my pals at the outdoor tables of my favorite brewpub against the panorama of Pusch Ridge, I feel that these wonders have been produced for me by the very forces I have given others money to suppress. What counts as creation or destruction, however, is based on values springing from ever-changing human subjectivities, with their subterranean desires and unexpressed ideologies. (Yesterday's sickies are today's cultural heroes—and vice versa.) The only plausible moment of pristine innocence must have been the microsecond before the Big Bang—and even that looks to be a bit suspect, if it was capable of producing a corrupted something out of a pristine nothing.

Despite the environmental setback of a president who comes off as a reincarnated Saudi oil prince, underwriting corporate greed while the well-being of American life is compromised by handouts to industry, we know from the Reagan era that a period of ecological reaction is destined to set in, and of course it is happening even now. But first, more damage than necessary will evidently be done, leaving more to be corrected afterward. Gas mileage, air-conditioning, seasonal energy efficiency ratios, alternate energy subsidies, etc., have all had recent setbacks from Republican rapacity and short-sightedness (not that the pusillanimous Democrats are so much better). But human life requires limitations to survive urban and suburban development, and once we are personally developed enough to become bourgeois, esthetic needs kick in as well, which increase the desire for limitations. Valuing nature is a middle-class enterprise, and even those putative despisers of American middle-class life who claimed to be rescuing the world for the rest of us, like Bernadine Dohrn, Sara Jane Olson, and Ted Kaczynski, turn out, in one way or another, to be products of bourgeois amenities. It's only after a full stomach has been assured that we are suddenly open to a whole spectrum of salvific epiphanies, not the least of which is the fantastic realization that the spotted owl, *c'est moi*.

Rattlesnakes, Shipwrecks, and Terra Incognita in Palos Verdes Peninsula

Stacy Warren

Rattlesnake

The rattlesnake must be dead by now. The old man with the gun must be dead. I know for a fact that the bushes growing wild, with the yellow flowers whose pungent smell nauseated me every spring, are dead: years later I saw with my own eyes that they were ripped out and replaced with symmetrical rows of decorative palm trees. The magnificent empty field of my youth—of all the local children's youths—is, for all intents and purposes, dead. It has been flattened, paved, and replaced with luxury homes priced so high that families with young children could not afford them anyway.

Photograph by Stacy Warren

Today, I can locate it in space. Palos Verdes Peninsula is on the ocean about twenty miles southwest of Los Angeles. I can tell you of its eighteen miles of rugged coastline, its elevation reaching from sea level to 1480 feet up steep, rocky cliffs. I can explain how it formed through successive marine terracing. I can recite the history, beginning with the abrupt transition from seafaring Native Americans to Mexican cattle and sheep ranchers in the mid-nineteenth century. I know of the early twentieth-century Japanese tenant farmers who hired Mexican labourers to grow garden vegetables for the L.A. market, and how, in the 1920s, the famed Olmsted Brothers showed up to design an elite planned community that would become a showpiece of the urban-planning literature. I also know from my own childhood memories that this so-called showpiece of community planning set into motion the trajectory that would lead to the luxury homes, dysfunctional families, and rampant consumerism that I recall growing up in the 1970s—the very conditions that ultimately killed my empty field.

I lived in Palos Verdes between the ages of nine and thirteen. After my family moved away, I rarely returned. I lived many other places and ultimately became a geographer, maybe in part because to this day I remain haunted by this landscape of childhood. I need to talk about Palos Verdes, the rattlesnake, the old man, and the empty field. And to begin there, I first need to tell you about the ideas of frontiers and terra incognita. There is also a shipwreck, but that comes later.

Frontier

In 1893—at the Chicago World's Fair, of all place—the historian Frederick Jackson Turner made a bold pronouncement. According to just-released 1890 Census figures, America's "Wild West" frontier was officially closed, a victim of rapid population growth. Turner declared that the United States had lost the terra incognita that had been heart and soul of the formation of the American identity. Without the frontier, the country would never be the same. Ever since then, geographers have paid especial attention to the frontier, both as concept and as place. I suspect it comes naturally as an intellectual category because it is also an emotional category that emerges in childhood: the mysterious topographies and unidentified flora and fauna of a child's frontier spaces, truly terrae incognitae. The childhood frontier need not be distant and unreachable; instead, children physically move them as animistic people might move through an enchanted forest in direct contact with their gods.

When I lived in Palos Verdes, every time I walked through the empty field I would see, touch, hear, and smell my terra incognita. The unknown was not a place unreachable and unconquerable because technology, or intervening political powers, prohibited it. Rather, the mysterious, intangible nature of my frontier was an active decision—a refusal to name, to categorize, to rationalize. Today, as a practising geographer, I am surprised by how little I formally "knew" about that field. I can't name the type of rocks found there or the soil. I don't recall the tree species or general elevation contours. Most vexing, I still can't identify that noxious yellow-flowering bush. As a child, though, I felt little need to label everything with scientific terms. What was important to me about that landscape came from an entirely different realm of knowledge.

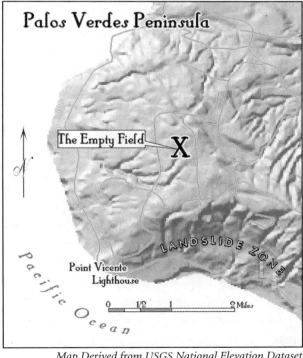

Map Derived from USGS National Elevation Dataset

I inserted myself into my frontier through pure experience, and revelled in it. All the children growing up in Palos Verdes in the early 1970s must have. We knew, for example, with absolute certainty, that an old man lived in a shack off in the distance, regardless of the fact that no one had ever actually seen him. We also knew with equal certainty that he had a shotgun and would shoot at any child who came too close. On the day of the rattlesnake, I was crossing the field on my way home from school. It was spring, which meant a daily dilemma: did I follow the beaten path across the center of the field and risk a shotgun blast? Or did I skirt the edges of the field and hold my breath past the nauseating yellow bushes? It was a shotgun-blast day, and I moved quickly through the middle of the field to get beyond where I knew the old man's gaze must reach. Suddenly two boys stood in my path. They motioned me to silence and pointed to nearby rocks. "It's a rattlesnake," they whispered. We all knew rattlesnakes lived there, exactly as we knew all about the old man, but we also knew that no one was actually supposed to see one. It's just part of their practical joke, I thought knowingly, and, ignoring the two boys, started to walk away. Then I heard it. It was unquestionably a rattlesnake; without ever hearing one before, I knew. I peeked around the rocks with great trepidation, painfully aware of how little I knew about real rattlesnake behaviour. Would it leap into the air to bite me? Would I die? And there it was, coiled up, rattling its tail at us, looking rather uncomfortable and miserable, as if hoping we would just go away and leave it alone. My entire personal relationship with the rattlesnake lasted less than two seconds; I fled immediately. I've never forgotten it.

 The odd thing now is that in spite of the importance of that rattlesnake to my terra incognita, I have no recollection of what the warning rattle sounded like—was it a low-pitched rattle? High-pitched? Rapid? Slow? The snake's existence in that field was presumed eternal. Because it would always be there, there was no need to organize or rationalize. I gave it no name, no classification beyond the imagery of pure experience. Yet now that it is irretrievable, I find I have no specific memories of the rattle. And I never did find out the name of that horrid yellow-flowering bush. Now I smell all bushes with yellow flowers; I ask people; no one knows. One of my life's minor missions was to return to that field with a botany book. When I finally did, there were no entries for palm trees in straitjackets.

Photograph by Bruce Perry

Shipwreck

Nature created the Palos Verdes Peninsula, with massive uplifts that caused the marine terracing in such quick succession that there was little time for lower terraces to erode. The individual benches are clearly visible, forces of pure nature, each one climbing higher toward the sky. Eventually, at the higher elevations, the terraces give way to hilly terrain punctuated by open spaces such as my empty field. These upper reaches of the peninsula are my next vantage point. Looking around me, I re-evaluate the contours of my childhood landscape from the perspective of an academic. I discover that terrae incognitae don't survive well into adulthood. From here, I can see how my childhood field fits into a wider topography—a few acres of relatively flat land, which drops off about one hundred feet on its eastern edge to a shopping center, and to the west gradually tapers downward into a trademark curved-street network of suburban tract housing. Now I can attach names to geographic features, and now I can interpret them in their environmental and cultural contexts. In fact, I can't avoid doing so; I know too much.

From the top of Palos Verdes Peninsula, I am compelled to retrace a path back down to the ocean but this time with the critical eye that knowledge gives. Beginning at my childhood field, I realize it was never empty or untouched by human hands. Today, my landscape memories are inhabited by the ghosts of the numerous herds of sheep who in the nineteenth century nibbled the hills down to the low brush we played around as children. During World War II, the flat expanses of my empty field led a mysterious double life as one-hundred-feet-tall radio towers sent coded messages to the Pacific theatre. Descending terrace by terrace down to the ocean, the ghosts of human modification are everywhere. The dramatic, steep barrancos so much a part of the rugged coastal personality are also relatively new, again the results of nineteenth-century overgrazing. Where I see hints of calcareous adobe soil on the more gently sloping terraces, I imagine the stooped figures of Japanese sharecroppers growing tomatoes and cucumbers for the Los Angeles market. I descend a bit further into the photogenic lower terraces. Their irregular profiles are nothing more than the effect of the construction of Crenshaw Boulevard in the 1950s, when tons of landfill inadvisably brought in atop of an ancient landslide literally pressed the lower reaches back to the site of their genesis, forty feet here, one hundred feet there. Now I've reached sea level and come to one of my favorite natural features as a child, the shipwreck of an old freighter. Back then I did not question that the shipwreck belonged in the same category as rattlesnakes, yellow bushes, and marine terraces. The hulk held all the glory, terror, and mystery of a natural formation. I never asked how it got there, or where it had been before, or why it wrecked. I didn't need to; it simply was there.

Now, today, the freighter is something else entirely—a symbolic trajectory, the final stopping place of human modification, the shipwreck of humanity. Its presence at the base of Palos Verdes Peninsula seems the logical conclusion to the sheep, farmers, and bulldozers witnessed on the path downward. My adult self is, after all, a geographer; I've been trained to read the semiotic codes. I can't resist deconstructing the freighter's mysteries. Who was on it, and what cargo did it carry? Did its wreck contribute to the abnormally high levels of PCB and DDT later found to be contaminating the Palos Verdes Shelf roughly one mile offshore? And what was that freighter doing on the wrong side of San Pedro Harbor in the first place? Most likely it was not lost in the fog; I know now that the lighthouse at Point Vicente might sound only 152 hours per year—only 1 percent of the time. Personally, I think it was compelled there as if by a siren's song to fulfill its destiny as empty symbol—literally, the symbol of nothingness.

The mysterious freighter of my childhood is today industrial exoskeleton, rotted and tilting thirty degrees askew at the base of an irreparable chain of environmental events. Its sailors abandoned it at the edge of the water, and everything else has washed back out to sea. Fittingly, even this tribute to industrial prowess has had no better luck surviving intact down here than the prickly pears and tall grasses that once covered the upper reaches of the peninsula. But I can't forget that this object I now label disparagingly as the "shipwreck of humanity" was once also part of my terra incognita, and to this day I still—in spite of all my geographic education—feel no contradiction in having a shipwreck be part of my "natural" environment. I don't even really mind that sheep nibbled away my frontier one hundred years before I got there. The only moment that truly hurt was when the bulldozers came and took away my field forever.

From this shipwreck I learn that loss of natural habitat unmodified by human hands is not the only form of environmental loss; the loss of cultural habitat can be equally as profound. It is childhood mysteries like the shipwreck that shape identity and form a powerful, lasting sense of place—the interior one tied to a million memories of a million pure experiences. Geographers watch this process eagerly, as people's lived and shared environmental experiences become rooted in the landscape and bring identity to place and, simultaneously, bring place to our individual identities. Geographers have words for this: sense of place, *genius loci*, cultural landscape. I had always imagined that one day I would write a book about Palos Verdes' sense of place, to capture the power I felt as a child. But after the bulldozers and the decorator palm trees, I imagine instead I'll write a book about its loss.

Dances with Bears

Jim Dwyer

"Hey kids, come join the bear dance!" Children of all ages loved doing the bear dance and getting bear hugs. Parents loved having a few minutes to look at the Grand Canyon in peace while their kids played. Park Rangers loved the bear for imparting natural history, conservation, and tips for hiking and camping, and for encouraging kids to listen to the rangers. I loved it all, especially when the women returned my big bear hugs. Dang, but it got hot in that furry suit!

Most people who've ever seen the Canyon love it, but some find mighty funny ways to show that love, like taking rides in helicopters whose vibrations damage both the fragile environment and the ancient Indian ruins. Others see the Canyon as an unexploited resource and an expensive traffic obstacle between the mills in southern Utah and the potential uranium mines. Why not have uranium mines and mill tailings in a windy area near one of the nation's leading tourist attractions? Let's use public funds to enlarge those dusty Forest Service roads and do some real mining!

The uranium mine proposal spurred an Earth First! campaign in 1985. Around the campfires, it seemed that everybody had some idea what their own little group was planning to do, but nobody knew the whole plan. The only ones who seemed to *need* to know all the details were strangers we quickly identified as FBI moles. Their trimmed mustaches, new T-shirts, and need to know were obvious giveaways. Besides, that's how people smoke *tobacco*. Those brilliant spies, misdirected, all ended up in the wrong place the next day.

But that's jumping ahead. Some of us made little pilgrimages to Edward Abbey's campfire because we heard that he didn't have much time left. His subdued attitude sadly confirmed the rumors. Somehow he managed to hang in there long enough to finish *Hayduke Lives*, which partially accounted the anti-uranium mining campaign. Some of the fun of reading the book was discovering that many of the characters were thinly veiled versions of actual people: "Hey, that's Bob!" Others were composites: "Remember those German ecofeminists?"

One morning the Flagstaff group donned radiation suits and started "cleaning up a small uranium spill" just outside the Tusayan park entrance. Tourists faced extremely realistic Forest-Service-style signs as they approached. *Tusayan Uranium Mine Scenic Overlook, 1 Mile. Caution! Blowing Uranium Dust Next 26 Miles.* Then came the "spill." Some drivers simply turned around and sped south, while others drove on through, but many stopped to find out what was going on. We gave out flyers about the problem. Some were indifferent and a few were angry at us—even threatening violence—but most were angry at the mining companies and government for even considering mining in the area.

A few cops didn't have any luck dispersing us, so eventually everybody arrived: the Freddies, the County Sheriff, and the TV crews. We'd made our point and left before we could get busted. Now, I don't know the details and have a terrible memory for names, but somehow while all those cops were busy with us, some survey stakes got pulled in the proposed mining area. In the canyon itself, several ruins were covered with tarps. The chopper pilots didn't want to fly in for a peek anymore. Gosh, another coincidence.

I traded the radiation get-up for the bear suit and returned to the park. "Hey, kids, do bears like berries?"

"Yeah!"

"Do *you* like berries?" I passed a big box of berries around.

"Yeah!"

"Do you realize that I'm not a real bear and that you shouldn't come close to real bears in the wild?"

"Yeah!"

"Did you know that real bears need a whole lot of protected space to live safely?"

"Yeah!"

"Do you think bears would like uranium mines in the middle of their woods?"

"No!"

"Do you want uranium mines there?"

"No!"

"Yow! Let's do the bear dance!"

Fast forward eight years. . . .

Willow Lake, a lovely subalpine lake with a large peat bog, is just southeast of Lassen Park. I was lounging by the lake on a virtual waterbed of peat, reading Terry Tempest Williams' wonderfully erotic essay "Undressing the Bear," when two acquaintances appeared across the water. I paddled over and offered them a cold beer to break the hot afternoon. It surprised me that they turned it down. "Gotta keep the temple pure man," one nearly chanted as he revealed three small multicolored pills. "These were blessed by Jerry." Although I hadn't had psychedelics take me for years and haven't since, this seemed to be the right time and place.

Several hours later, after a spectacular sunset and full moonrise, I suggested a moonlight hike to a geyser less than three miles away. They declined, but I found the allure irresistible. About halfway to the geyser I saw what appeared to be a large porcupine walking across the top of a down log. As I approached it stood up and revealed itself—a bear cub. That's when I entered a total flow state and everything started moving in extreme slow motion. In an instant that seemed to go on forever, I spotted the mother bear about forty yards away on the other side. I turned on my halogen flashlight, blinded the mother bear by flashing it in the eyes, and waved the cub away. I backpedaled to the trail, shouting the first words that came to mind. The bears went one way and I the other, no harm done.

At the geyser I experienced an epiphany that turned into a poem as I hiked back to camp. When I wrote it down, I recognized it as a series of haiku.

Midnight Mist

Geyser spewing steam,
blowing pulsating white puffs
into ebony sky.

Wandering deer herd—
two doe, two fawns, one four-point
buck—appear through mist.

We eye each other,
the buck and I. Silently
we both hold our ground,

Look up at the moon.
As clouds of steam pass over
it becomes dimmer

then brightens again
in clear sky, then darkens in
rhythm with geyser,

ever seeming to
grow larger and then smaller
in a changing beat.

As the steam passes
a full spectrum of color:
Moondog rings the moon.

I am not alone
in fascination: the deer
all gaze heavenward.

Upon arriving in camp, I faced two related questions from my companions: "Why were you yelling so loud and who the hell is Terry Tempest Williams?" (The first words out of my mouth had been "Terry Tempest Williams!!! Terry Tempest Williams!!!")

I answered with another accidental haiku:

A junkyard moondog
howling through the clouds, chasing
heartsick thieves away.

"Is that you, Junkyard Moondog?"

"The *Reverend* Junkyard Moondog to you." A nom de plume was born.

Two years later, during a Terry Tempest Williams session at the Western Literature Association Conference in Salt Lake City, I recounted the Terry-bear story. Everybody got a big charge out of it, especially the fellow who turned out to be her husband. He asked me to repeat the story to her. We had a short but enjoyable conversation. She responded with a smile as bright as the northern lights. Occasionally I've been disappointed that my mental image of a writer was rather more impressive than the actual person, but Terry turned out to be every bit as brilliant, challenging, life-affirming, and complex as her work.

The next day I took the tram to the top of Alta and hiked over to a neighboring peak. I saw a big, bearish guy skiing in a black ski snowsuit and laughed at the thought of bears on skis. Then I asked myself, "What do bears like to do that I like, too?" Lacking food, I stripped and wallowed in my finest ursine style.

Suddenly, a snowboarder appeared, seemingly from out of nowhere. "Dude! Rad! But I'd put some sunscreen there if I were you."

I donned my shorts and asked if he had any spare water. When he removed the water from his jacket, a book fell out: Doug Peacock's *Grizzly Years*. Since it was clear that we already had at least two mutual interests, snow and bears, I thought I'd try for a third.

"Have you ever read anything by Ed Abbey?" I asked.

"*The Monkey Wrench Gang*."

"Did you know that some people think that Hayduke is based on Doug Peacock?"

"Get out! The person who turned me on to this book knew Abbey."

"Who's that?"

"Terry Tempest Williams."

So I told him the bear story, leaving us just one thing to do: the bear dance!

Finding the Forest: Citizen Activism in the Truckee River Watershed

Michael P. Branch

I started early, took my dog, and headed west out of Reno, snaking along the Truckee River, up through bare canyons filling with morning light, up through fog-banked meadows, up into the rich, silent forests of the Tahoe Basin. I was headed for a place I had learned to care something about, a particular patch of forest that has become a destination, the way home becomes a destination when you drive toward it from a faraway place.

I'd been at home in other forests: the mixed deciduous forests of my southern Appalachian boyhood, and again in the tangled cypress stands and shaded mangrove hammocks of the Everglades detour my life took as I made my way slowly out to the Great Basin. Still, the stately, genteel coniferous forest of the Tahoe Basin had always seemed like someone else's woods, its picturesque slopes and open stands and sculpted boulders too like an advertisement for property you couldn't afford, a ski chalet waiting to happen, a place that only existed for two weeks in August.

These are the limitations of vision we carry with us on our scattered pilgrimage through new landscapes. Like the cross-country drivers who burn blindly across the beauty of the Great Basin with a psychological discomfort only thinly veiled by esthetic disdain for the "nothing" they see there, I was vaguely suspicious of all these arrowy, soughing conifers, this clean and green postcard forest of the Tahoe Basin. Where were the humid, breathless recesses of the hazy Blue Ridge hollows or the moccasins and gators of the bromeliad-draped cypress sloughs? This breezy, bright, needle-carpeted cathedral of the Sierra was somehow too perfect. It looked like a place where a moonshiner couldn't hide. A place where things never died, and dead things wouldn't rot. A place you could admire but never love.

I had been a forest activist before, in other places, and had always been especially fond of watchdogging logging roads—those arteries of pain invading the healthy body of wild country—and helping to ensure necessary road closures and obliteration in the interest of forest health. As a lover of poetry as well as forests, my motto had been, "Two roads diverged in a yellow wood, and I got them both closed." But now it seemed time to learn something more about these Sierra forests, to get behind the postcard image and around to the cryptic message on the back and then to begin to decipher that message to see who loved whom, whether all was well, and to discover, if possible, what might be missing from my own way of inhabiting the Truckee River watershed.

Like rafting a river just before it's dammed, walking these shadowed forests was always poignant and heavy, for you walked in the shadow of the tractor and the saw—a long, chilling shadow you could feel as it fell across something that had reached evening before its time. On some days forest monitoring felt like little more than measuring the waning life of something that was inexorably doomed; on the other days, though, the woods felt as restorative and quiet and endless as they had seemed when I was a boy.

One afternoon in the summer of 1997, I had just come off a would-be logging unit in the North Shore Timber Sale with Rich Kentz of the League to Save Lake Tahoe. As we sat on the tailgate sweating, drinking water, and looking out over the blue, windy expanse of Lake Tahoe, Rich offered a brief explanation of forest history in the basin. During the Comstock era of the mid-to-late nineteenth century, much of the basin's forests had been severely cut. The ancient ponderosa and sugar pine were the first to go, providing the beams, braces, and scaffolds that would support the miles of mine shafts beneath then-booming Virginia City. A century later, increased human habitation of the basin has resulted in ambitious fire-control efforts, by which homes have been protected but the forest has become loaded with an unnaturally high concentration of dead and downed trees, thus exacerbating the risk of catastrophic fire.

In response to the fire danger caused by heavy fuel loading, the Forest Service strategy is to log these areas in order to thin dense forests (which are dense because the natural fire cycle has been interrupted) and eliminate the "ladder" fuels by which a low-intensity ground fire may climb to the canopy and escalate to an uncontrollable holocaust. While reasonable in principle, the reality of such salvage logging has often proven to be environmentally destructive. The economic reality in the basin is that the trees we need to remove in order to reduce the risk of fire—dead and downed trees, ladder fuels, jackstrawed ground fuels, saplings, and young trees, especially white fir—are precisely those with the least economic value. In order to make logging pay, it becomes necessary to cut some big trees, some green trees, and some pines as well as firs—an approach that results in a disturbed and less diverse forest of mostly younger trees. Add to this that logging results in soil compaction and erosion, destruction of wildlife habitat, threats to water quality, and sometimes residual slash that actually *increases* the danger of fire, and you can understand why forest activists and other citizens worry that the solution may be worse than the problem.

A tailgate is as good a place as any from which to begin the reformation of the world. Rich wrote a proposal in which he and the League would persuade the Forest Service to remove 240 acres from the planned North Shore Timber Sale. Instead, we would go in and take care of the forest ourselves. The Forest Service agreed, and the Twin Crags Forest Project was born. Working with the League and other volunteer groups, Rich worked out a treatment plan with the Forest Service, rounded up the necessary equipment, recruited a group of team leaders, and had a number of preparatory meetings, usually in the field, on the forest unit we would work. It was a special pleasure to collaborate throughout the year with the team leaders, a diverse and talented group of foresters, activists, prodigies, whistleblowers, smart-asses, and rednecks, many of who combined these various identities to particularly good effect. I enjoyed their companionship and passion for the forest, their idiosyncrasies and humor, and their thorniness and concern.

One year from the tailgate, August 29, 1998, we held the first Tahoe Forest Stewardship Day. That was a day of days. More than two hundred volunteers joined us to make real what had seemed possible only in dreams. We had birdwatchers, mountain bikers, boy scouts, activists, ROTC folks, students, and scientists working side by side to thin stands of saplings, eliminate ladder fuels, redistribute and chip ground fuels, re-grade landing zones, mitigate erosion, obliterate skid trails, and protect wildlife habitat. Patagonia and other area businesses pitched in with food, drinks, prizes, tools, and other kinds of support. Hydrologists, wildlife biologists, and fire ecologists were on hand to answer questions, to teach volunteers about forest ecology, and to add a crucial, educational component to the event. Forest Service staff worked with us to help us make good decisions about how to manage on-site work, and media coverage was supportive and informed. Participants worked hard, had fun, stayed safe, and felt good about their contribution to the forest. As we gathered at the foot of the mountain for a barbecue at the end of the day, there was little left to wish for.

Nearly a year has passed since that day, and in the intervening months I have continued to return to this patch of forest. Although, as today, I often bring only my dog for company, the presence of those hundreds of volunteers is with me now whenever I visit. This place has a new layer of story, and it is not the same old story of cut-and-run exploitation. The poignancy of that long shadow is gone from one place at least, and the forest is back to doing what it does best: healing itself and sustaining everything around it. It is good medicine to me to be in such a place in all seasons. Happily, the success of last year's efforts has resulted in an expansion of our project, and this year we will work on two additional sites. The Forest Service, at first hesitant, is now enthusiastic about the Twin Crags Project as a prototype for collaborative, community-based forest management.

It is a fair hike into the unit, several miles with moderate elevation gain, and I am content to be walking without human company. At the lower elevations I am struck by the glowing beauty of incense cedars, their thick, aromatic bark reddish, twisted, and dotted with acorn-sized holes. Chartreuse moss adorns the north sides of the trees in symmetrical rings. A pair of Stellar's jays flash indigo, chasing and squawking through the lower branches of the surrounding trees. The trail becomes steeper and I pause to rest. With long, silent wingbeats a Clark's nutcracker glides, a noiseless arc of black and gray, into the crown of a large tree below me. The quiet here comes as such a surprise that it occurs to me that silence, so commonly driven from our daily lives, is yet another balm that the forest produces, which it stores and holds, as it does water.

As I rise and turn up the mountain, I notice a sliver of silver moon hanging, inverted, in the blue western sky. I am now above the cedars, in the thick of the Jeffrey forest, and fast rising into mixed conifer stands in which the fir, both red and white, become larger as I climb. On the way I pass the telltale signs of previous logging operations: skid trails that disappear like scars into the forest, landing zones perhaps a decade old that remain lunar and nearly unvegetated, half-burned piles of slash that someone decided weren't worth loading, a tangle of scrapped skid cable—the industrial version of a beer can left behind. Nevertheless, this forest is recovering, beginning to stabilize, to heal itself, to hold water again, to clean the air, to be home to animals, and to replenish its soils. Even the worst of the areas, those stripped to bare, mineral soil, have at last begun to show signs of life, as tiny sprigs of lupine struggle to colonize this miniature wasteland in the heart of the forest. As I reach the crest of the hill, I see what I have repressed all along: the familiar yellow sign, facing me, that announces yet another timber sale. Nearly everything I have walked through today will be purchased and logged, even as the land struggles to recover from having been logged so recently. It's the old story again: the world is for sale to the highest bidder, and the long shadow is not easily escaped.

I look down at the dusty toes of my boots and push past the sign and into the unit of forest that is not for sale—the forest that is the home of the Twin Crags Forest Project. A magical line is crossed as I step from a world being auctioned into a world of trees that will be allowed to stand their ground. Here is a lovely, dominant red fir forest that won't come crashing down, a recovering forest floor that won't be skidded, and healing soils that won't be scraped away by the blade of a tractor. Here is a preserved deadfall that will be left undisturbed and will make fine winter quarters for a denning critter. Goshawk, California spotted owl, and the many other birds we have heard and seen here will return to their familiar perches and nests high in the forest. The black bear, whose berry-spangled scat marks the trail today, will have a familiar place to browse and loll. We will all have a place worth returning to, a place that can heal and in healing can make the world feel right again.

For me, this forest project has meant a way of finding home in the

Truckee River watershed. It is a slow process, this work of finding home. We must see a place through its days and seasons, watch as the wheels within wheels of our home ecosystems turn to the forces of weather and time. We must also see these landscapes through the slowly deposited layers of our own perception and sensibility: one day alone, one day with a friend, and one day with a group of volunteers; one day through sadness, one day through contemplative stillness, and one day through joy. The land gets inside us, and we begin to think of its beauty as something related to what we feel is best in us and its destruction as something that affects us deeply, personally, and viscerally.

Maybe, if enough of us continue to care about our home in this watershed, I'll be able to return to this patch of forest during some autumn in the autumn of my life, when the skid trails will be invisible, and the telltale orange markings on the trunks will have slowly faded away, leaving a healthy forest of straight, clean, furrowed bark, standing tall and ready to catch the flakes of the first snow.

Right in My Back Yard, or Confessions of a RIMBY

T. Louise Freeman-Toole

Environmentalists are often accused of being obstructionists who react to every perceived threat to their neighborhood with the cry, Not In My Back Yard! The next time someone calls me a NIMBY, I'm going to say I prefer to think of myself as a RIMBY. After years of protesting against unwelcome changes in my part of the world, I've learned to concentrate instead on positive, compensatory actions I can take Right In My Back Yard.

My back yard is a unique and little-known part of the Northwest, called the Palouse. Unlike the wet western side of Washington State, this hilly region straddling the border with Idaho is arid and largely treeless. Once the Palouse Prairie contained a wealth of plant life—an estimated thirty-five different species per square meter; now it is a fertile farming area whose extraordinarily rich soil has been dumbed down to grow just one thing: wheat. Strictly speaking, local "growers" also raise peas, lentils, and garbanzo beans, but the vast majority of the land is planted in wheat. Wheat occupies a central place in the culture of the area, both symbolically—as the staff of life that feeds a hungry world—and monetarily—as the economic lifeblood of the region.

In the spring, the Palouse is as uniformly green and as neatly tended as a golf course. Later in the season, there might be a swathe of brilliant yellow rapeseed on a distant hillside, but there is never anything blue visible. Yet the reminiscences of early residents, both the native Palouse Indians who lent their names to this area and the farmers who supplanted them, all mention—in a wistful, so-beautiful-it-hurts kind of way—the hills of endless blue. Every spring, they said, camas flowers bloomed as far as the eye could see, and each dip and hollow seemed to hold a pond reflecting the sky. Herds of fine appaloosas bred by the Palouse Indians roamed the hills, getting fat on the rich grass.

The Palouse Prairie remained largely unchanged until the development of the self-leveling combine made it possible to harvest crops on the steep hills. Over the years, more and more land was plowed until today less than 1 percent of the prairie remains. Pockets of native grass, wild roses, and balsamroot survive on the few hills too steep for even modern combines, and prairie remnants can still be found in the small cemeteries that dot hilltops around the Palouse. These peaceful family graveyards offer some of the best views of the area, and these vantage points remind me that even without most of its native vegetation, the Palouse is still beautiful. It's a cozy, fairy-tale landscape dotted with church-spired villages and clumps of trees tucked among the hills; surely, any moment now, a friendly giant will come striding over the far hill, cupping a little farmer in his hand. (It's no coincidence that most Green Giant peas are grown on the Palouse.)

It doesn't take much effort to envision a tiny thirty-two-horse team pulling a heavy, old-fashioned combine up one of the hills. It takes real imagination, however, as well as some knowledge of native flora, to construct a picture of the Palouse as it existed for centuries before the arrival of the plow.

Several years ago, when interested locals came together to form an organization devoted to conservation and restoration of the Palouse Prairie, they found that most members had only a hazy idea of what the area had looked like before white settlement.

Few had ever seen a patch of native prairie or would have recognized one if they'd stumbled across it. Some, like me, had a romantic but rather vague vision of pretty green hills with appaloosas running across them. Several members from local farming families remembered their great-grandparents' stories about spending long summer days wandering the hills picking buttercups, larkspur, bluebells, and lupines. The scientists in the group talked about protecting rare species like Jessica's aster and Palouse goldenweed.

My idea of a prairie had been shaped by the *Little House on the Prairie* books, and I imagined the hills covered with lush tall grass waving in the wind. But the more informed members of the group told me that the Palouse Prairie was very different from the tallgrass prairie of the Midwest.

I soon learned that in our dry western climate, native grasses grow in low-growing bunches scattered across a crusted soil. Drought-tolerant flowers clung to the hillsides, some of them bearing waxy or hairy leaves that help them to retain moisture. A variety of fruit-bearing shrubs supported a healthy population of sharp-tailed grouse and other game birds. The Palouse Prairie was a whole different animal—not as simple, but even more interesting—than I'd thought. I got to see this for myself on the first of the group's regular field trips.

We carpooled out to a rocky hilltop on Paradise Ridge where one of the founding members of the Palouse Praire Foundation had built a house overlooking a patch of undisturbed prairie. For many members, it was their first glimpse of the Palouse Prairie. We walked gingerly along the hillside, stepping carefully around each precious flower. It was a low-key landscape that invited you to slow down and look carefully. What seemed at first glance to be a slope covered by a nondescript, heath-like vegetation became on closer look a richly textured landscape full of a variety of shapes, colors, and smells. A few people got down on their knees to take a closer look at a prairie smoke, a delicate pink flower with a silvery plumed seed head. There was an occasional cry of delight as someone recognized a rare plant—a broadfruit mariposa or a Palouse thistle—that they'd seen before only in photographs. Everyone was smiling; some were almost giddy. I felt transported, too, because the landscape felt both novel and familiar to me, like meeting a long-lost relative with whom I felt instantly at home. It wasn't the first time I'd experienced this kind of group euphoria after a close encounter in the wild, but it had always been in response to something huge and spectacular like spring break-up on the Yukon or a humpbacked whale surfacing just off the bow. There was nothing showy about this postage-stamp prairie with its gentle colors and ground-hugging profile, and yet it had elicited a similar sense of wonder and connection to the natural world, made all the more poignant because this was, after all, our home.

After this initial outing, we went on to visit other prairie remnants on farms on both sides of the Washington-Idaho border. We traipsed through the yard of a local optometrist who'd replaced his lawn with grass and flowers native to the Palouse, and we toured the community garden where volunteers were attempting to establish a patch of bunchgrass. Seeing the enormous time and energy that had been invested in these still-struggling plots made me think twice about attempting such a restoration project in my own yard. One factor that makes raising native grasses particularly difficult is that many of the native grasses require some kind of scarification—passing through a bird's guts, or being subjected to freezing temperatures—before they will sprout. I reluctantly concluded that, as someone with little gardening experience, I had small chance of success. My back yard was huge and overgrown and I didn't even know the names of any of the plants or trees. I was just grateful that they survived without getting much attention from me.

But as I went on more field trips, seeing a variety of prairie remnants and restoration efforts all across the Palouse, the sense of familiarity I had experienced on Paradise Ridge grew. I learned to identify many local wildflowers, shrubs, and trees, and this knowledge enabled me to look at my yard with new eyes. I realized, somewhat to my embarrassment, that I didn't need to plant native species in my back yard—someone had already done it for me. The prickly bush where sparrows congregate in winter I now recognized as Oregon grape. What I had always called a pine tree was actually a Douglas fir, a favorite of locals who cut their own Christmas trees in the national forest. The birch that turns a lovely gold in the fall was, more specifically, a river birch, like those that grace the banks of the shadowy St. Joe. The bush that releases a sweet perfume in late summer is mock orange, a relative of the syringa that lines the route Lewis and Clark followed from Montana. The witchy looking trees along the fence that have defeated all my efforts to cut them back are wild plums. I even discovered a serviceberry bush crowded against the fence by an overgrown lilac. Early settlers in the area had made jam from the berries, so I was surprised to find that they had little flavor. It was obvious to me now: virtually every shrub lining the perimeter of the yard was native to this area.

I didn't know who had landscaped the yard with plants from the Palouse Prairie, but I felt a sense of connection to him or her. In addition to feeling thankful for the anonymous gardener's efforts, I felt a responsibility to carry on what had been started perhaps as long as forty years ago.

I cut back the lilac to let the birds reach the dark blue serviceberries. I planted another mock orange bush in place of a shabby forsythia alongside the driveway.

Next to the porch, I put in a red-twig dogwood whose burnt red bark is handsome even when bare of leaves. I pruned the mountain ash and it produced a bumper crop of red-orange berries, which hung on the tree all winter, providing food for a flock of cedar waxwings and a family of deer. One summer I planted hardy local blanketflowers on a steep section of the yard where grass had never flourished. I let the wild plums grow.

I know now that the Palouse Prairie will never be what it was. I'll never see scores of appaloosas running across the hills through shimmering lakes of camas flowers, scattering blue petals in the grass. But I can find solace for the loss in the most unlikely of places—right in my back yard. I can sit and breathe in the summery Coppertone smell of mock orange, watch little brown birds pluck berries from any of a dozen shrubs, and know that RIMBYs like me can help ensure that at least part of the Palouse Prairie will endure.

Section II
Challenges

Wedded to the Cause

Robert Schnelle

A few days of hiking in Washington's Cascade mountains doesn't feel like much of an environmental slugfest or a way of preserving anything, but it gives you heart. When my wife and I became "grove guardians" of several square miles in Wenatchee National Forest, we used our commitment as an escape, playing hooky in the backcountry under the pretext of saving trees. By September morning light, Lori and I did ground-truthing in wooded bottomlands. We penciled lines on our map for the unacknowledged motorcycle trails and the overgrown skid roads. We marked undisturbed sections of forest with a felt-tipped pen, taking care to amend our species count and our records of old-growth timber stands. But all of this felt to us like a married couple's retreat. Afternoons, as we loped along a serpentine backbone of ridges, we savored the internal hum that comes from sharing silence broken only by ambient sounds. Waist-deep tangles of thimbleberry, cedar fronds lilting beneath the mist of a waterfall, bobcat tracks in the dust of an outcrop at six thousand feet—these were the benchmarks of our so-called labor as we paused to spread our lunch on a nurse log, holding hands as we rested. I scaled a hemlock tree and felt the wind rushing eastward from Snoqualmie Pass. Lori photographed the Stuart Range framed with yellow-tinged boughs of enormous larches. Days in the forest couched us in mosses while we dreamed at night. Asleep, we seemed to hear the scrunch of footfalls in corn lilies.

Later on, after posting our report on a website for the Cascades Conservation Partnership, we did nothing further but wait for the levers of organized advocacy to trip the wheels of government. Wendell Berry has complained about the "pinhole vision" of the environmental movement. But sometimes narrowness pays off. October brought news of appropriations from Congress that would protect local forests and help maintain a regional wildlife corridor. The hope is that enough patches of habitable land might be linked to ensure the future of Northwest plants and animals. Owls, bears, and cougars depend on room to roam, to say nothing of the forest-dwelling Sasquatch, who may not exist, but whose fiction (if that's what it is) voices the spirit of the woods in uncanny hoots and howls by starlight.

Our own role in Bigfoot preservation, I know, was tenuous. The project we took on felt like a brief love affair, which was heightened by the release it offered from everyday life. Apart from writing one among dozens of other field reports, Lori and I had no part in the hearings, the faxings, the phone calls, and the strategy sessions that resulted, for instance, in there being enough shade for trout in Taneum Creek, enough acres of shaggy trees to allow the red vole—which never touches ground within its lifespan—to find itself a home.

Two years on, I think of the time we contributed as a vacation, an ecotour minus the air travel and haute cuisine—some sacrifice, right? But taking credit is beside the point. With the White House occupied by the worst environmental president ever, we'd be foolish to dwell on success, whatever the scale of our commitment. If this is true for Bobby Kennedy and the big guns, so too for those of us working at the grass roots. Here in Kittitas County, Washington, we've recently learned that our elected board of commissioners is poised to sell off a stretch of the Yakima River's floodplain—"surplus" public land they call it—for conversion to the Stone Resort and Evangelical Conference Center. No doubt these wetlands will beef up tax revenues once the sale goes through, but the hidden cost is 12 percent of what habitat remains for the river's salmon fry.

So, although Lori and I would rather be hiking, we have decided to renew our political efforts. In the world most of us live in, there is no free love: doing right by nature means talking into a microphone in front of people who hold you in contempt; it means writing letters to the editor and noticing how pompous or goofy your words look when printed; it means crossing the boundaries of e-mail etiquette to wheedle support from office mates. In the case of an admired friend, commitment leads you to spend 350 bucks for a newspaper ad, later to discover that your carefully worded arguments have appeared on the comics page. And then there are the meetings and the phone calls, the false hopes, the tedium, and the frayed nerves that attend the activist's calling.

No wonder so many people avoid causes. I, too, dislike them. Terry Tempest Williams has called the environmental movement "a lover's embrace," but it must have been people like Julia "Butterfly" Hill she had in mind. Hill is the young Californian who perched for two years in the highest branches of a redwood tree and saved a swath of ancient forest for her trouble. The anchorites, and those early Christian mystics who lived for decades on top of stone pillars, would have approved the urge that drives some of nature's friends to risk their safety. In several western states, goons have burned down the houses of conservationists; other greens have lost their lives in shady circumstances. Confined to an eight-by-twelve-foot platform, Julia Hill was buzzed by men in helicopters shouting obscenities.

For Lori and me, activism is more like clerical work—a lot of nitpicking and botheration. We put down the coffee cup and write a check. To propagate protest we click on SEND. It is strange that with the world going up in smoke, our efforts on behalf of life should feel so tepid. We bridle our emotions in order to appear sane in the eyes of decision makers, but we know it is they and their patrons, the powerbrokers and exploiters, who carry on in a state of delusion. Inwardly, we nod at Henry Thoreau, who asked, "What madness possessed me that I behaved so *well*?" This he wrote after harboring a fugitive slave and going to jail to protest the Mexican War. My idea of acting up is to circulate petitions.

Yet it is clear to anybody who reads *Walden* that Thoreau was nobody's dad. Like Rachel Carson, Peter Matthiessen, and many other lone-wolf activists succeeding him, Thoreau followed his calling beyond the family circle. This leads one to ponder whether his books would ever have seen print had Ellen Seawall replied, in answer to his proposal, "Yes: yes, Henry, I will." If middle-class environmentalists are meek, maybe our wives and husbands prefer lying next to someone who doesn't shout epithets in their sleep, a person for whom "fronting the essential facts of life" includes coaching in the junior soccer league. Perhaps we are reticent in order to protect our kids, who attend public schools with the kids of agribusinessmen and snowmobile enthusiasts. Under the influence of my fourth-grade son I only daydream about monkey wrenching, and I make it a point to speak to our neighbors, property-rights vigilantes though they are. It's not that I refrain from taking Erik to the woods with me. He knows the cheer of a campfire and the press of roots beneath his backbone. He can explain why burgers hasten the death of tropical forests. Even so, I censor what I say around my son—I've learned to soft-pedal ideology. For children who worry about cargo pants and the amperage of their gameboys, outdoor time is precious. We shouldn't box in their capacity for solace.

In the end, my environmental movement is like a marriage. With all of monogamy's steadfast satisfactions and petty complaints, it's a habit of being that sustains the future through present sacrifice. It involves the sort of work that no one congratulates you for (nor should they), though you may feel you are living your life on a Möebius strip. For if we suppose, as some have maintained, that marriage is not a word but a sentence, we can face up to our covenant with the Earth. Though its crises keep licking our ankles like a flooding tide, environmentalism is the effort to *countenance* nature, as we countenance each other. It is our worldly wise way of declaring, with St. Augustine's god, "Love means I want you to be." In our modest campaigns I am blessed by my wife's company, the breadth of commitment we share extending to the land we live in, I hope. Days we seem to have worked for the good, or times when something precious was lost, are among the ties that bind us to matters ever so much greater than ourselves. To revive our passions we will still go hiking. We will sound creek bottoms with our walking sticks, bask on lakeshore gravel, and notice calypso orchids rising among the ferns.

Then, back in town, we will wonder where the salmon can shelter when the floodplain is smothered by putting greens. Then and there we will realize that together, our promises renewed and our paths worn smooth by others before us, we have to put our shoulders to the wheel.

The River Home

Amanda Gordon

I stand on the riverbank in mud-soaked boots, brown chestwaders, and a green slicker waiting for the run of the Pacific Northwest coho salmon. Here on the West Fork Smith River, I gather data for the Oregon Department of Fish and Wildlife to record the life history of the coho salmon and, in my own mind, to find out why they are disappearing from Oregon's watersheds.

Inside the trailer where I'm staying this fall to gather data, dawn pours through the windows and creeps over my sleeping eyes. I sit up in bed with a sudden jolt—I've got to get the salmon out of the trap. They like to travel more distance during the night when there is less chance of being discovered by predators. They are animals that live by the moon. I scramble out of my sleeping bag and jump outside, shocked by the cold bite of morning air. Throwing on my waders and boots, I grab the net, cradle, clipboard, and tagging gun, and run down to the trap as fast as I can. I move with a sense of urgency, for I cannot stand the thought of the fish that are unable to get upstream.

Just as I expected. Three dark shadows, quivering in the light. Two are ahead at the upstream gate. I can only begin to imagine the slow death creeping through their panicked bodies. The third is swimming near the back of the trap. I lower the ladder down into the trap and climb in. This frightens them; I am frightened too. After attaching the handle to the net, I take a deep breath, and make the quickest lunge I can with the net. Faster than I can react, the fish slips away. It swims between the net and me. The water forced against my legs by the whip of its tail almost knocks me over. I stand still for a moment; I get the net ready and dive again. My arms, back, and legs immediately tighten, and I lift the violently jerking sides of the net's frame. The weight of the thrashing fish is more than I can handle. I reach down and take hold of the net on the sides with both of my hands; maybe this way I can keep my body from being thrown completely into the water by the weight of the fighting fish. The limits of my physical strength are up against years of struggle and fight. I am scared of this salmon's force, of her body throwing herself against the net, trying to get out. Fins slapping the water, head thrashing, teeth grabbing the net, all in a frantic, primal energy. This is a dangerous business for both of us.

I hold the fish there for a few seconds until she tires of beating her body about. Her side is painted in brilliant watercolors. Bright pink brushes her lateral line, and the rest of her body is silver—a sign of the camouflage she has developed since entering the river again. Her scales are coated by a thin epidermis, which in turn is protected by a mucousy covering, her seal of life. I must be careful not to damage this fine protective coating; a careless touch could kill her. Her upper jaw points in a V-shape down over the front of her lower jaw. Her slightly gaping mouth reveals rows of needle-like teeth. But it is her eyes that draw my attention: black pearls washed by the river's resistance to her drive to get upstream. She's spent the last two years migrating in the coastal waters eating marine invertebrates, crab larvae, smelt, sand lance, herring, rockfish, squid, and members of her own family—chum and pink salmon. She was probably only four inches long when she first entered the ocean and doubled in size in only the first few weeks of her ocean life.

Holding the handle of the net between my tightened legs, I reach into the water. With one hand I grab the peduncle—where the body meets the caudal fin—and try to flip the fish over. This way, she will be completely disoriented, and I will have control over her body. My hand struggles to wrap tightened fingers and palm around wet, slick, tight muscle. The fish senses my tightened grasp and flicks her tail, sending my arm flying up in the air. She is stronger than I am. There is no muscle in my body that matches the one long, tight muscle of hers. This salmon is longer than my own leg, thirty pounds I imagine. I try again, this time grabbing the peduncle and flipping the fish over quickly. My other arm slips under her, and my trembling hand cups the midsection of her body to support her weight. My hand slowly moves down the length of her body. Fingers stroking river muscle. It is coho, muscle, fight, cold, life, death, and I. I can see scars running up and down her body, wounds from the sea, from making her way upstream, or wounds from the trap, from her initial attempts to leap out of the enclosure with anticipation. My hand finds the top of her head and rests there. Trying not to lose control, I lift the thirty pounds out of the water. She immediately begins to twitch for breath in the terrible air. My chest and arm muscles tighten. I lower her quickly into the cradle before she can get away, and I immediately feel her body relax in my grasp. Her head is under a rubber shade, a calming darkness. I quickly take her measurements: thirty-eight inches, the largest I have seen and a record size for the coho.

I lift her up by the peduncle to determine her sex, already certain that this salmon is female. Only the females can move with this quickness and strength. Her body is on the verge of bursting with the need to release her eggs in the gravel beds of her birth. Just to be certain, I push along the length of her underbelly to see if any sperm will squeeze out. My hand widens as it passes over a swollen body; nothing appears but pink flesh. Her belly is full of a thousand dreams that she will take home. No male will fight like the female. They are bigger, stronger, carrying the hope for the future.

The West Fork Smith River curves its way through clear-cuts. The watershed eventually drains a portion of Oregon's southern Coast Range into the Umpqua River, just below the ridge, and out to the Pacific Ocean. The federal government claims more than half this land, managed by the Forest Service and the Bureau of Land Management. Land management here has meant dam building, road construction, clear-cutting, timber harvest, and the removal of the large woody debris that is home to the salmon: it protects them from the storms, the predators, and the strong current. Logging was the leading industry in Oregon for many years, and the first logging done here was along the rivers. That way, the logs could be easily transported downriver for milling. Convenience is everything. Not only are the trees cleared in doing this, but so are the riparian zones.

Our project that studies the coho, the Coastal Salmon Restoration Initiative, receives funding from the Bureau of Land Management and Sport Fish Restoration funds. It used to be funded in part by the logging industry. However, when the coho were federally listed as threatened with the help of our project, the timber companies withdrew much of their support, for if the fish are threatened logging regulations must be tightened. This paradox is killing the coho.

Looking down at the thirty-eight-inch female, I think about her looking for a shelter here below this mess. She needs the protection of all those trees that were cut down way above the riverbed. I wonder if she will find her natal spawning ground. If this was her home, it's gone now. A mudslide that wouldn't have occurred if the trees had not been cut has buried what might have been her gravel bed, and the pools are washed out. I think of the sediment that's not being held to the mountain by the trees. If she does find a place to lay her eggs, and if the sediment doesn't smother them, it will choke her fry as they try to emerge.

I pull down the tagging gun and push its needle under the scales just below the salmon's dorsal fin. In my desire to tell her that I am only trying to save her family, I drive in the gun. Please don't break any bones, don't shatter her vertebrae—or worse yet, damage her nerve cord. If I aim wrong I could paralyze her. Shoot the gun, twist, and pull out. No blood this time. I do it again on the other side, quickly. She is now a number to be recorded in history. But I don't want to see her here anymore. My hand grabs her peduncle one last time, and I pull her out of the cradle. She is tired now, too tired to fight. Her gills pump water in and out, moving at an unconscious rhythm with the current. My other hand carries her head and I heave her over the gate, out of the trap into the recovery pool. She doesn't swim out right away, but lingers. Her yellow tags stick out of her sides—my permanent mark on her body. I turn and reach for my clipboard to record the data. Important numbers for later, for trying to solve the mystery of their lives. I look up again to see how she is doing, but she is gone. River water flows between my legs holding me back, and I whisper goodbye.

Civil Disobedience and the Arctic National Wildlife Refuge

Carolyn Kremers

When I was twenty, a California jury tried me for participating in a nonviolent march on a national day of protest against the Vietnam War. I was convicted and served my sentence, and since then I have worked peacefully for social change—as a writer, a public school music and English teacher, and, for the past decade, a professor of literature and creative writing. Over the past two years, however, I have considered participating in nonviolent civil disobedience again. And I hope that I am not alone.

Having lived in Alaska for sixteen years, I have driven all eight hundred miles of the Alaska pipeline, and I've toured the oil fields at Prudhoe Bay. Two summers ago, my former husband and I hiked and river-kayaked more than sixty miles in the Arctic Refuge, where, for nine days, we saw no other human being. It tears my heart to imagine the Refuge coastal plain—two thousand acres of it or more—invaded with drill pads, runways, service roads, pipelines, gravel pits, air traffic, barges, trucks, noise, kitchens, and dormitories, all of which will be necessary if oil and gas development are allowed in the Refuge.

However, my opposition to drilling in the coastal plain is not just based on disturbing images. Nor is it based simply on the logic and statistics that many journalists, politicians, and activists are currently debating. For more than a decade I have observed the struggle over the Arctic Refuge, and I have concluded that it is also important—indeed, essential—to consider the moral issues.

It seems to me that the moral issues surrounding the Arctic Refuge do not center on animals—the caribou, polar bears, and birds that migrate there to have their young, or the musk ox, wolves, wolverine, moose, Arctic and red foxes, white Dall sheep, and black and brown bears that visit or live there.

Neither do the moral issues center on people—the Gwich'in Athabaskan Indians of Alaska and Canada, who live along the migratory route of the Porcupine Caribou Herd; or the residents of Alaska's North Slope Borough—mostly Inupiat Eskimos—including the people of Kaktovik, the only village located within the Refuge; or the other residents of Alaska, most of whom, like me, rely on jobs for cash income and are pleased to receive annual payments from the Alaska Permanent Fund Dividend program, which was originally established to help distribute profits from the oil developments at Prudhoe Bay.

I do not believe that the moral issues regarding drilling in the Refuge center around the citizens of the United States, either, even though they "own" the million and a half acres of the Arctic Refuge coastal plain, which is part of the nineteen-million-acre area designated a refuge by the Alaska National Interest Lands Conservation Act of 1980. Nor do the issues center around the citizens of the world, who have lost to development much of their land, air, and water, and who may view the Arctic Refuge as one of the last remote, intact ecosystems in the world.

The moral issues, it seems to me, center on two unlegislatible human traits: wisdom and greed.

The people of the United States comprise only 1 percent of the world's population, yet we consume 25 percent of the planet's resources. That includes more than seventeen million barrels of oil per day. Scientists and others have told us repeatedly that fossil fuels cannot last indefinitely. The technology for conserving fossil fuels and for developing alternative energy sources already exists, but what is required now is public awareness and support. *Now*, not twenty or fifty years from now.

The people of the United States and Alaska, however, choose to be greedy and to demand *more*. More oil for our deodorant and credit cards and fake wood and Ziploc bags with plastic zipper pulls and televisions and video games and electric kitchen appliances and food preservatives and unbreakable dishes and artificial turf; pantyhose, lipstick, nail polish, fan belts, snow tires, dashboards, motorcycle helmets, snowmachines, four-wheelers, cruise ships, jet planes, and on and on.

In fact, according to figures released by the U.S. Centers for Disease

Control and Prevention, in the year 2000 nearly forty million American adults were obese—one in five—and more than half of Americans, or 56.4 percent, were overweight. These rates are increasing in both genders and among all population groups, and child obesity is also on the rise, with teenagers almost three times as likely to be overweight as twenty years ago.

Researchers suggest that this increase in obese and overweight adults and children in America is likely due to high-calorie, high-fat eating habits coupled with decreasing rates of physical activity. It seems to me that linking weight with diet and exercise must now be taken a step further, to linking weight in America with oil. Many products made from oil allow and, I believe, encourage unhealthy diets and sedentary lifestyles.

I wish we might ask ourselves which uses of oil are essential to our health and well-being and which are not. I wish we might choose to give up some items made from oil and choose to use others mindfully and less often, with more awareness of where they come from, how long they last, and why we value them. I wish we might make changes in our lifestyles and workstyles and seek better ways to coexist with the rest of the planet.

But I know that some Americans prefer to allow drilling in the Arctic Refuge, on the chance that oil will be discovered there and will be economically recoverable—neither of which is guaranteed—and knowing that it will likely take a decade to bring any oil production in the Refuge on-line. If, by then, oil prices prove high enough to support the cost of production and transport, and if the Refuge coastal plain is able to produce an average of 225,000 barrels of oil per day for forty years—as President Bush and the most optimistic USGS figures suggest—that will only amount to about 3 percent of what the United States currently consumes daily.

Is it worth it?

✦ ✦ ✦

Here in Alaska, I have learned to pay attention to where airplanes land, what trucks and trains carry, how work crews are housed, where insulated boots come from—and long underwear, work gloves, welding masks, lumber, sheet metal, forklifts, bulldozers, beds, blankets, eggs, bacon, and hot coffee.

If Congress votes to allow oil and gas development in the Refuge—this year or next, five years from now or ten—I hope I will have the courage to choose to help block the delivery of those goods and services.

It is daunting to talk about some of this, particularly in light of the attacks on the United States on September eleventh, and all that has happened since. I do not consider myself a troublesome person. I'm a woman, recently turned fifty-one, who writes and teaches. I love my country, fiercely. I'd rather not spend my time and energy—and possibly risk my life or health—disobeying the law. I'd rather not contemplate fines, jail, and another criminal record. Yet I feel that, if Congress votes to allow drilling on the coastal plain, I must choose to participate in and advocate nonviolent civil disobedience, right here in Fairbanks, the center of services and supplies for the North Slope, even if I am one among only a few.

The soul of the Arctic Refuge coastal plain is unique, profound, and irreplaceable, and to allow penetration of that—for forty years of saran wrap and thirsty automobiles—is to participate in the pinnacle of greed.

Maybe, though, things won't come to this. Maybe the members of Congress will realize that they are at a pivotal moment in the evolution of humankind and our awareness of our place within the natural world—a world which includes polar bears and terrorists, wings and fire. Maybe some will understand, as Gwich'in Athabaskan Sarah James expresses it: *There are places that shouldn't be disturbed for anything.* Maybe enough U.S. lawmakers will choose wisdom and courage, and the Refuge will be spared.

If not, I hope that all Americans who care about the planet will search their souls and determine what they can do next to confront this insidious greed, which threatens to destroy us more secretly and thoroughly than anything we have seen.

Buffalo Field Campaign

Dan Brister

I sit on the flats above the Madison Bluffs burning sage for the sixty-nine buffalo. Looking toward the trap, I feel the helplessness freshen in my gut. It has been a year to the day, and tears rise as I remember. They confined us as they carried out their goals, arresting anyone who so much as stepped foot on the Forest Service road, restricting our movement to a small patch of ground beside the cattle guard and a barbed wire fence. More than twenty law enforcement officers from the Department of Livestock (DOL), Fish, Wildlife, and Parks, the Montana Highway Patrol, the Gallatin County Sheriff's Department, and the U.S. Forest Service collaborated to keep us far away and ineffectual.

A breeze blows up from the river and carries a thread of smoke from the sage bundle out across the flats toward the trap. I envision, one by one, the faces of my six friends who were hauled off to jail that day as the cops made the area secure for the kill. I can almost see the brown and flowing mass of bison, almost feel the rumble of their hooves in the earth. Closing my eyes I picture the smirking DOL agents riding horses and snowmobiles, hounding the animals with cracker rounds and shouting "Yee-Haauw!" as they stuff sixty-nine wild bison into the tiny steel trap.

We had been with those bison all winter, skiing out in the frosty dawn to find them so we would be ready when the livestock agents came to kill. We had been successful on many occasions, shepherding the buffalo into heavy cover where the DOL agents, on their snowmobiles, couldn't find them. We'd worked with local landowners who let us post our *Bison Safe Zone* signs on their property boundaries. Earlier in the winter we'd built and maintained a blockade on Forest Road 610—the only access to the site of the Horse Butte trap—preventing the DOL from killing for nearly two months. But on that Wednesday, the 14th of April, 1999, the DOL came to kill, and there was nothing we could do but shout and pray from the cattleguard or go to jail.

Horse Butte lies at the heart of the bison-killing controversy. Only a few miles from the border of Yellowstone National Park, the peninsula supplies some of the richest, most fertile habitat in the Greater Yellowstone Ecosystem. It provides lush winter range for the Yellowstone bison herd, as well as myriad other species like bald and golden eagles, white pelicans, trumpeter swans, sandhill cranes, great blue herons, coyotes, gray wolves, grizzly bears, black bears, moose, deer, and elk. While their primary target is the buffalo, the Montana DOL affects every species in the area.

The Department of Livestock has a helicopter to haze buffalo back into the park. Flying a few feet above the treetops and shooting cracker rounds—loud firecrackers fired from shotguns—the DOL agents in their helicopters are deafening and disruptive. We have video footage of a flock of three hundred trumpeter swans being flushed by the chopper and other footage of an elk caught in a bison hazing operation hooking its leg as it attempts to jump a barbed wire fence and landing on its head. Because there are three known bald eagle nests on Horse Butte, much of the area is closed to human activity under the Endangered Species Act. Yet the DOL repeatedly violates the eagle closure, both flying their chopper above it and riding their horses inside it—all of which we've captured on videotape.

These operations are conducted at an enormous expense to taxpayers to protect the interests of a few livestock producers. One hundred and forty-two cow/calf pairs graze Forest Service lands in the Horse Butte area, bringing the U.S. Treasury eight hundred dollars a year. The current bison management plan, in effect until 2015, is slated to cost the federal and state governments between two and three million dollars a year.

The state blames brucellosis for the slaughter, saying the presence of bison in Montana poses a grave economic threat to the state's livestock industry. Buffalo originally contracted brucellosis—a European livestock disease that can cause cattle to abort their first calf—from cows. Montana argues that it has spent millions of dollars to eradicate the disease from its cattle in order to be certified as "brucellosis free" by the United States Department of Agriculture (USDA). This certification allows livestock producers to export cattle without testing for brucellosis. Federal regulations, however, don't justify the state's response. The USDA has admitted that the presence of infected bison in the state does not jeopardize Montana's status as brucellosis-free.

The risk of wild bison transmitting brucellosis to cattle is so remote that not a single such occurrence has ever been documented. South of Yellowstone, in Grand Teton National Park, where bison and livestock have coexisted for the past forty years—and where a greater percentage of the bison herd is infected—a case of brucellosis has never appeared in cattle. In its 1998 study, *Brucellosis in the Greater Yellowstone Ecosystem*, the National Academy of Sciences concluded, "Under natural conditions, the risk of transmission from bison to cattle is very low."

Fear of disease transmission isn't Montana's true reason for killing buffalo. Brucellosis is present in many species—elk, deer, moose, coyotes, wolves, bears, and bison. There are twenty times more elk than bison in the Yellowstone ecosystem, and elk have transmitted brucellosis to livestock. Yet elk, which bring Montana more than eleven million dollars in the form of hunting licenses and permits, are allowed to roam freely between the park and Montana, unmolested by the DOL. West of the park near Horse Butte, where all the killing has taken place in recent winters, bison and cattle do not come in contact with one another. Because they can not survive the severe winters, cattle only range from June to October, when the buffalo occupy their summer range inside Yellowstone.

Brucellosis is a smoke screen for sentiment that runs much deeper. Viewing public support for the 1995 reintroduction of wolves into the Greater Yellowstone Ecosystem as a weathervane, the livestock industry fears a shift in public perception regarding federal lands management. What's really at stake is the image, in the public eye, of bison re-inhabiting any of their former range outside the park and competing with cattle for grass. An area game warden, who asked to remain anonymous, spells it out: "If the public gets used to the idea that bison, like elk and deer, should be free to roam on federal lands, then it may lead to a reduction in the amount of public lands forage allotted to livestock. That's what the ranchers really fear." In the days before cattle, Montana supported millions of buffalo. By 1900 the vast herds, which had occupied much of North America and numbered as many as fifty million bison, had been reduced to a single herd of fewer than two dozen individuals in Yellowstone. On their former range, much of which is now public land, stand millions of cattle. If buffalo are allowed on public lands outside the park, they will compete with cattle for forage and set a bad example, an example that Montana's livestock industry won't tolerate.

Yellowstone National Park is the only place in America where wild bison were not eradicated. The miraculous recovery of this herd, which grew from twenty-three individuals in 1902 to nearly four thousand by the mid-1990s, is widely cited as America's most successful conservation achievement. Yet it is a tenuous achievement at best. Yellowstone buffalo are being killed in greater numbers today than at any time in the last 120 years. During the winter of 1997 more than half the herd was lost. A slaughter instigated by Montana's livestock industry and compounded by an unusually severe winter reduced bison numbers in the park from thirty-five hundred to less than fifteen hundred animals in a few months. For the buffalo, 1997 was the bloodiest year since the nineteenth century.

The group I work with, the Buffalo Field Campaign, formed in response to the slaughter of 1997 and has been running patrols ever since. Ours is the only group working in the field every day to stop the slaughter of the Yellowstone buffalo and advocate for their protection. Since our inception in the fall of 1997, more than a thousand people from around the world have joined us on our front-lines patrols to protect the buffalo. We are grassroots activists dedicated to informing the world of the plight of the buffalo and organizing opposition to stop the slaughter. The backbone of the Campaign is our daily documenting of every move made against the buffalo. Volunteers perform acts of civil disobedience to keep the animals safe. Because we are with them when they are out of the park and in danger, we are able to serve as a strong voice for the buffalo.

On that April day in 1999, hemmed in beside the barbed wire fence and the cattle guard, we didn't feel very strong. Some of us shouted, "Run! Run! Run!" at the buffalo, hoping they would bypass the trap and drop down the bluffs to the safety of the Madison River—and they almost did. At the last minute the herd hesitated on the brink of the bluff, allowing the agents to circle around them and chase them back, into the open mouth of the trap.

Others sat quietly and watched the scene unfold, praying for the buffalo. The sixty-nine buffalo, and another two captured that day in a different trap, spent the next two days and nights in confinement, before most were shipped to slaughter. Four died from injuries sustained while confined in the small trap.

    I think back on all this with a year's perspective, during a mild winter when no buffalo have been killed. Kneeling in the peaceful meadow burning sage for the buffalo we couldn't help, I wonder when the killing will end for good and what we can do to make that day come sooner. As I look up from this thought, a group of buffalo cows and calves appears on the rim of the bluff and walks toward me on the flats. I hold still. Ten feet from me the herd parts, half passing to my east and half to my north. I am in the center of a circle of thirty bison. Sage smoke drifts into the nose of a pregnant cow who lifts her head and stares at me. I sit in amazement, surrounded, and draw strength for the work that lies ahead.

Dear Mom and Dad—Send Food

Chuck Pezeshki

What do you do for spring break if you're an activist? You go to camp. And in the Northern Rockies, it's Buffalo Field Camp, to stand between one of the last bison herds left on the planet and Montana Department of Livestock (DOL) and U.S. Forest Service officials trying to haze and/or shoot the buffalo that have wandered outside Yellowstone Park.

For most of the year, Yellowstone's wild bison herd is happy to stay put in the park, grazing. But in late winter they get hungry, causing some to run for the western border to the bottomlands of the Madison River, just outside the park. It is here that they run afoul of Montana DOL.

The state of Montana has steadfastly maintained that brucellosis, a disease found in all sorts of wildlife, can be transmitted to domestic cattle from the bison, which can cause domestic cows to abort their calves. Never mind that there's never been a documented case of transmission to cattle from bison in the wild or that 50 percent of all elk in the Yellowstone ecosystem also carry brucellosis. But unfortunately for the buffalo, Montana DOL puts bullets behind their words. They killed 1083 bison that came out of the park during the harsh winter of 1996-97, with another eighteen hundred dying through prevention of migration, reducing the Yellowstone herd by two-thirds. Such heavy-handed management impresses large agribusinessmen across the West—the people that really run things in the state—and facilitates interstate livestock transfer. They don't want those pesky bison out of that park. The line—actually a clear-cut line—is sharply drawn, defining the square of the park, running all four directions. If only buffalo had some sense of plane geometry.

The issue is really a suburban one. The town of West Yellowstone, immediately adjacent and the gateway to the park, resembles more a parking lot than a quaint little mountain town. In the last twenty years, it has become *the* snowmobile destination of the world, with up to ten thousand machines revving their motors outside hotels from November to March. And surrounding West Yellowstone are the subdivided communities of Hebgen Lake. It's what happens when people get a home where the buffalo roam, a jumble of suburbia and the high lonesome bluffs where buffalo used to have their primary calving ground. There used to be a broad valley here, filled with the meandering river, and wetlands that made green meadows late into even the driest year. That was before the Hebgen Lake dam. Then Quake Lake formed in 1959 by a huge landslide—it stretches almost to the base of the dam for Hebgen Lake. The best land was flooded, and that has cramped everyone's style. The houses are on a peninsula or pushed up against the mountains. But the irascible buffalo keep looking for a way out—whether to walk out on the Horse Butte peninsula or make their way along the north shore, to the plains beyond. Combine that with the Montana DOL buffalo-capture facilities, and you have a unique version of buffalo hell.

I'm a middle-aged activist, and when I go on spring break, some of the basic elements have changed. First, it's no longer just me. Everyone comes— my wife Kelley, Grandma, and our two toddlers, Braden and Conor. Second, we go to camp in that symbol of middle-class suburbia—a minivan. There is holy purpose behind this adventure. I am a professional photographer, and I'm shooting stock photos for the Buffalo Field Campaign (BFC), an Earth First!-style effort dedicated to keeping the migrating bison safe from the Montana DOL. They do this by tracking both the bison and the DOL. Sentries are posted in beat-up cars around this rural/suburban matrix, and BFC volunteers stand ready to screw up any attempts to herd the bison into any one of the capture facilities.

The BFC runs out of a cabin on the north side of the lake. The cabin, though not small—four rooms plus a kitchen and bathroom—is packed with people. Two bedrooms are filled with about thirty bunks. Dinners are eaten communally—lentil stew, rice, and bread. The first meal tastes good, but by the end of winter, the diet just doesn't cut it. The calories and protein are there but lots of necessary vitamins and minerals aren't. Among the activists that are staying for the whole season, there's a strange smell of malnutrition wafting on the breeze.

At the nightly meeting, which runs more like a battle patrol recap than a social-change campaign, Pete Leusch, one of the directors, asks for a guide to accompany me. Megan, a sweet soft-spoken woman in her early twenties, volunteers.

◆ ◆ ◆

Early in the morning Megan and I load up to go buffalo hunting. "We only need one good picture," I tell her. "That's the thing to remember." The goal of this trip is to burn some negatives that can be blown up billboard-size. The Indigo Girls are major BFC supporters, and the request has gone out for backdrop photos. Megan and I encounter our grazing brown brethren almost immediately. I follow the buffalo around, snapping away, lolling my head to act as unthreatening and ridiculous as possible. The buffalo aren't interested in posing. I chase them around with the view camera, a black dark cloth over my head.

◆ ◆ ◆

Later, as we walk into the motel lobby, Megan slows down, then freezes. She's staring at the complimentary continental breakfast bar, with its row of juice machines lined up on the Formica countertop. Megan's been at camp all winter, and she's got the look of a lizard watching a fly, an intense stare of vitamin deficiency. "Juice?" I inquire. "Please," she says, breathlessly. I leave her with four full glasses of orange and apple nectar. "Can you manage this while I go up to inform the family of our whereabouts?" With one glass already in her hand, she smiles.

On the third day Megan and I tramp into a BFC patrol camp in the willows along one of the river's long, sinuous curves. We are headed for Houdini's Meadow, named after a big bull that successfully eluded a DOL joint helicopter/snowmobile hazing attack last year. The sun is out, and two big bulls graze in the field. We shoot keeper shots, the expansive view of the Madisons behind wintering bison.

After packing up and hiking out, Megan recommends a family hike to see the trumpeter swans that winter in an unfrozen pool where the Madison hits Hebgen Lake, just east of the Horse Butte capture facility. But as soon as we arrive, a woman comes running, calling for help. Two sheriff's cars, a U.S. Forest Service rig, and a truck hauling three snowmobiles are entering the capture facility. Braden, sensing the tension, starts screaming, "Papa, don't go—don't go!"

By the time I run up the road, law enforcement has already blocked off access to Horse Butte. DOL agents are racing their snowmobiles around three bison. The bison are giving the agents a run for their money, darting up the adjacent butte in an area closed to snowmobiles for protection of nesting bald eagles. The protesters, confined to the road, scream at the cops, and cheer every time the buffalo run up the slope.

Then all hell breaks loose. The buffalo burst through a fence, toward the protestors, snow machines following. The protesters increase the cacophony, then rush between the bison and the machines. Undeterred, the agents dogleg around the group. The bison take off for the river, one stumbling and rolling in the snow, the whole gang running down the road after the snowmobiles. Polish cavalry and Nazi tanks—this must have been what it was like—an unnerving exhibit of the overwhelming advantage of the violent energy of hydrocarbon society.

The buffalo turn again, back toward the pens, herded by the snowmobiles. A young man called Lumpy jumps up, into the closed area and toward the bison. A foul-tempered sheriff tackles him, yells "You're under arrest!" and binds his wrists with plastic restraints. Lumpy is off to earn his fingerprinting merit badge. I turn to another protester, Anna. "This type of thing really doesn't bring out the best in law enforcement," I remark. She nods her head. "It doesn't quite bring out the best in us either."

With the three bison in the pens, the crowd flees—word has it that there are three more bison near the Cougar Creek capture facility, and the BFC volunteers aren't going to give them up without a fight. One of the leaders, a long-time friend, Mike Bowersox, hands me the videotapes from the event. "No one's going to arrest you," he remarks.

Bumping down the road in the back of a dilapidated pickup, all I can think is that this is one hell of a way to spend vacation. If this activism is just a hobby, I need to find something more relaxing to do.

When I arrive at the BFC cabin on my next-to-the-last day at camp,

Pete and Mike have bad news. Seven bison have already been captured, and a blockade has been set up on the Horse Butte Road. I hike cross-country in knee-deep, slushy snow, crossing the road at the edge of the police tape strung around the blockade. On the road proper, a young activist, Hawaii Pete, is twenty-five feet off the ground, suspended in a basket from a tripod lashed together from trimmed lodgepole pine trunks. He's sporting a lockbox, a metal contraption that binds his arms around two of the tripod poles.

The tripod has been studded with nails and wrapped with sharp wire to prevent arresting officers from using chainsaws. The protesters catcall at police to "play nice" with the obstructer. At first, it seems preposterous—one does nonviolent confrontation with the intent of getting beat up. The idea is that the outrage precipitates change, and telling the cops to be nice is against the rules. Then Hawaii Pete reveals that he's a hemophiliac, and if he gets bruised, he's going to die—fast.

Two cops are standing in the bucket of a lift truck scratching their heads. At one point, the cops want to chop the poles out from the bottom or saw off the top. I put on my engineering professor hat, tell the tree cops that this is a very bad idea, and remind them that I'd be happy to testify in a civil suit after Hawaii Pete is dead.

The cops finally manage to get Hawaii Pete in the bucket and start talking deal if he unlocks. Instead, he demands they free the seven buffalo already captured today. The peanut gallery cheers. Allison, Hawaii Pete's support person, the only person present that can give hemophiliacs the clotting serum, runs around outside the tape, begging for access. One officer responds, "I could care less." And then it is over. Pete's in custody, but the delay has spared the bison a daily roundup.

That night back at the motel, my children romp on the bed as I watch the nightly news. The bison story comes in third, and the press gets it wrong.

Three months later, I receive the BFC newsletter. It is upbeat—great progress is being made, people are signing petitions, volunteers are tabling in Yellowstone, and the buffalo are that much closer to salvation. Send money—that too is included in the missive. Oh yes—and food. Anything: old survival rations from Y2K, leftovers from the long winter. No one's too picky at Buffalo Field Camp. Please?

Looking Down from Hale-Bopp

Colin Chisholm

Every year for a decade now my friend, Tom, and I meet somewhere in the desert. Our goal—and it is, increasingly, a very difficult one—is to find a place where we won't see a human being for a week. Going off tips from friends or map guesswork, we've managed to spend lots of time alone in the canyons. Our quest for solitude is addictive, a need to get away from the ever-increasing speed of our everyday lives.

This year we return to a remote area we visited many years ago. Back then there was no trailhead, no trail, and no cairns. We'd parked the car beside the highway and vanished into the desert. This year we arrive late on a March day, appalled to find that a trailhead has been bulldozed, as well as the beginnings of a road. A bulldozer rests nearby. At once a vacuous feeling I recognize but can't name slithers into my gut. Tom shakes his head; no words are necessary.

Despite this transgression, we stick with the plan; we have come too far to turn around. Our plan is to hike up one canyon, over the mesa top and down into another. Let's call them Billy and Bob Canyons. Both have long, sinuous narrows and require rock climbing and a few rappels. We shoulder our packs and hike east up Billy. Potholes brim with water, primrose bloom, and cottonwood leaves glow translucent in the twilight. At nightfall we find a slickrock shelf where we cook dinner to an owl's soft, reedy hooting.

Later, we lie back and watch the sky, emblazoned by the Hale-Bopp comet, which comes around once every ten thousand years. I huddle in my bag, feeling like a child on Christmas morning. All those stars, each with its own story. I try to fathom what the Earth was like the last time Hale-Bopp came around. I sleep so well in the desert.

Canyon wrens wake us in the morning, their descending trills echoing off canyon walls. The desert bestows rituals like small gifts; time and again they surprise. Waking to canyon wrens is one of these. Sleeping on slickrock, because it is clean and smooth and soft, the search for water, and silence are also rituals. To fall into these rituals is to rest and to heal. They are rituals desert people have always followed, as far back as Hale-Bopp's last journey, maybe farther. Who are we to change them?

As soon as the sun hits us we begin hiking. The canyon constricts, until my elbows bump the frigid, fluted walls, colored like the flesh of salmon. Goose bumps pepper my arms. Deep in Billy's guts, where sun may never warm the walls, we swim up to our necks through longer and deeper stretches of water. My clicking teeth echo.

Shivering and laughing hysterically at once, we come to the end, where a climb through an angled slot leads to a patch of sunshine; we sprawl naked like drenched lizards. I am euphoric that a place like this still exists away from the human throng. But the trailhead? The bulldozers and the new road? Suddenly my joy is tainted by a deep sadness; I am mourning the loss of this place even while I wade through its profundity.

I am no stranger to this mourning, nor are most people. Many of us have lost the wild places of our childhoods. Before I'd turned eighteen my childhood meadow had been bulldozed for a golf course. I began to believe that any place I loved would be plowed over or paved under in the name of progress. All around us wild places slip away, and we're in a kind of frenzy to know them before they're gone. Thus the boom in outdoor recreation and the rabid growth of the new west, where funhogs rule the land.

Night births a full eclipse of the moon. It's been twenty years since I saw one, when my little brother woke me with his cries, afraid of what he thought was God's eyeball in the sky. I fall asleep quickly, waking only once to Hale-Bopp, a waterfall of light spilling onto the silken land.

On the fourth day we cross a freshly graded dirt road. I ask Tom where it's going and he reminds me that most roads lead nowhere. From high on a mesa I stare into the vast desert, and everywhere these nowhere roads crisscross the land like so many lines on a map. A spume of dust rises from one like a harbinger of the teeming masses waiting for the light to turn green just beyond the horizon.

Bob is much like Billy. Lush gardens spill from cracks in leviathan boulders. Water seeps from springs, potholes glimmer in the shadows, and bobcat tracks wind leisurely through the mud. We stumble upon an Anasazi ruin. Thumbprints of the makers cover the walls like a sea of tiny waves. At night, coyote calls spiral from the canyon rim.

On the last day we find a Snickers wrapper, a bloody Band-Aid, a Nike sneaker, and blooms of toilet paper. Boot prints abound. The spell is broken; for the first time all week I can't hear the little things. My brain floods with white noise. I smell the bulldozer before I see it, tearing its way into the desert.

People are coming, whether I like it or not.

In the long run, perhaps none of this matters. The desert thrives on a geological time scale, while humans think mostly in terms of our short lives. Billy and Bob Canyons will be here in a thousand years, relatively untouched. The roads and our bones will blow as sand into the desert.

Tom and I make the mistake of driving back through Moab, where Jeeps seethe around us. I want to be angry at these people, but what I feel is shame, because I am not so different. My complicity is apparent in my own desires, only slightly different from their own.

I roll down the window and breath in the sweet smell of sage. Only when we are twenty miles out of town do I calm. We stop to camp, and looking up at the night sky I find Hale-Bopp. I stand there, silent, wondering what this place will be like the next time Hale-Bopp comes around.

The Toll Road

Lee Schweninger

From the window of the bus on a trip to the airport, I watched a woman as she walked through a field on what I assumed to be her farm. The old farmhouse stood in a distant patch of cottonwoods on the rise above the creek. The woman—who looked to be in her seventies, with silver hair tied back in a tight bun—moved deliberately along a double-track path through the wheat grass and blue grama, through the buffalo grass, through what appeared to be relatively undisturbed prairie land. She wore a white blouse and blue jeans, and she sported a straw sun hat, which hung from a string around her neck. This woman paused often in her walk; she stopped, not to rest or to catch her breath, it seemed, but to sweep her hand across the stems and tassels of the seed pods of these grasses, as if offering benediction.

She appeared careful, thoughtful; she seemed to know the secrets of the grasses as one might know an intimate's secrets. Indeed, she traced the grasses as one might caress a lover, her stroke light and lingering. As she dipped her hand into the grass, it was as if she touched me, as if I could feel the soft flesh of her palm on my own body.

All this I saw and felt in a moment as the bus angled along U. S. 36 just east of the foothills of the Rocky Mountain Front Range between Denver and Boulder. This highway goes by many names: 36, the Denver-Boulder highway, simply the Turnpike, or even the Pike. But most often, and even now several decades after the toll booths have been removed from under the Wadsworth Boulevard overpass, some of us old-timers still refer to it as the Toll Road. There were never any revolving spikes blocking the road, but at Broomfield, the halfway point, there used to be a set of toll booths tucked under the bridge. There used to be a very literal and direct cost associated with driving this highway, twenty-five cents each way.

Whenever I return to Colorado to visit family along the Front Range,

I ride along the Toll Road on the bus toward the foothills or from the foothills. Formerly prairie or farmland, much—maybe even most—of the land along the highway is now devoted to suburbs and industry; huge industrial parks and housing tracts of clapboard sprawl across what was formerly, during my childhood and even young adulthood, rolling hills of mixed grass prairie, what was formerly wild, or land devoted to farming and horse grazing.

Prairie grass is resilient. Its dense web of roots survives repeated fires and constant grazing, and we could still learn from that resilience. We could learn to hold onto the soil, protect the soil, and give back to the soil. We could learn to use the wind for energy as grass uses the wind to pollinate and carry seeds. We could learn to conserve water as grass blades know to close tightly around the moisture from rain and dew, as grass knows to conserve that most scarce and ultimately sacred resource. Rain comes to the prairie only in fits and starts, and prairie grass holds on and preserves that water, treating it as the precious manna it is.

The prairie offers a lesson in resilience. We could walk amidst the switch and bluestem and redtop and vernal and reed canary and buffalo and orchard and grama and tobosa and purple needle. Amidst the grasses and flowers, aster and rose and prairie star and chickweed and clover and ragged robin and thistle and teasel and crazyweed and locoweed and Queen Ann's lace and queen of the prairie. We could walk with butterflies and moths, sulphur and skipper, and swallowtail and viceroy and monarch and the regal fritillary, and with other insects, mantises and bees and beetles. And we could walk amidst birds, hawks and vultures, golden eagles and kestrels, grouse and crane and the burrowing owl, kingbird and swallow, sparrow and lark bunting. There are hundreds of species, thousands of species, to every swamp, swell, and swale.

On the prairie we can witness renewal, we can conjecture life, dream of our roots and of the people in our blood. The prairie is vast enough for our dreams and still larger than our imagining. Surely the people are grass.

Conversion of the prairie along the Front Range in Colorado has been especially intense in the past decade. Contractors gouge the land, fill the valleys in, lay the hills low, and cover them with asphalt for spur traffic.

Between 1988 and 1996, traffic volume on this highway increased by 45 percent, almost doubled; some sixty-five thousand vehicles a day travel this road. This number has been predicted to double again in the next twenty years: 130,000 vehicles a day, ninety vehicles a minute, a car or truck or bus every three-quarters of a second, twenty-four hours a day.

The transmutation of the land flanking the Toll Road and the consequent traffic is indicative of the metastasis along the Colorado Front Range in general, the entire one 130-mile length, from south of Colorado Springs to north of Fort Collins. Thousands upon thousands of acres of farmlands are for sale. Available. Acre by acre, section by section, these prairie, crop, and grazing lands are cracked, leveled, and bounded. Hills made low, drainages forced underground into concrete culverts; in their stead rise up houses, businesses, streets, parking lots, automobiles, and buses, like the one I so often ride.

I was flooded by these thoughts as I watched that woman through tinted glass; she straightened, drew her hand away from the grass, as if from my flesh, shaded her eyes, and glanced westward. She looked across the prairie toward the mountains that rise abruptly out of that prairie, mountains capped with the pinks and oranges of sunset clouds, mountains from which I'd just come. The moment passed quickly; an instant later the woman herself was out of sight; the RTD airporter pulled me away to the east.

In the few minutes immediately after the bus passed, that woman became for me a symbol. I burdened her with my assumption that she was somehow saying farewell, as I was saying farewell, to these acres of prairie that had been her home, that had been my home. I imagined that these grasses had surrounded the home of her parents as they had surrounded the home of my parents. That the house behind her had been the home of her enterprising children, now moved away, like the home of my parents, whose children were now moved away.

Perhaps I was being whimsical, romantic. Grass is not a symbol, not an idea. I know that. Grass is real; grass is flesh and bone. Real seeds, pollinated, break from stocks and germinate, become grass topped or lined with seeds. I know grass is real, so perhaps I transferred onto this woman my own emotion, my own feelings of regret for the loss of my homeland, the loss of these grasslands. I projected onto her my own personal loss. Perhaps I fused my own geographical distance with the literal loss of the prairie. Maybe I cherished the irrational expectation that we can honor and preserve the grasses, that politicians, business people, and construction bosses will share my values, that human beings can grow and change, can learn from the grass. Perhaps I silently acknowledged that by pursuing my own career that called me across the country, I gave up and lost this prairie; perhaps I foreswore any rights to the land. This woman in her field piqued me, reminded me that I have forsaken the land, shirked my responsibility. She pointed directly at me, as if to say, "There are great costs for abandoning the land you live on."

As her words echoed inside the bus, inside my head, I craned my neck to look for her, but through the window I could barely even see the mountains behind us. I felt a wave of nostalgia for the grasses and mountains beneath the skies that I had grown up with; I felt a homesickness for the wind and earth that molded me. Prairie sky. Prairie wind. Prairie grass.

I admit it: I imposed onto that woman my own feelings, feelings that may well have had nothing to do with the feelings or mood of the woman walking on what might not even have been her field. But I take her being there and her movement as an emblem; she paused in parting to dip her hand into the smooth stem of grass and seed pod. I take it as an emblem of the loss of all prairie. I take her delicate farewell, her hand along a grass stem, to be an emblem for my own loss, and of humanity's loss, not only of the prairie in particular, but of the wild itself.

On my next trip along that highway, just a few months later, I remembered to look for her, remembered to scan the landscape for the woman whose hand touched the prairie whose grass was flesh. She wasn't there of course. But to my amazement, neither were the prairie grasses where she had walked. I was shocked to see that the land she had traversed had been cleared of all grass whatsoever; in just a few months a row of tract houses had suddenly emerged, cracker-box houses fronted by concrete driveways, asphalt streets, and curbs without sidewalks. Raw and ominous around the new houses, where had grown the prairie, lay brown clay like cracked scar tissue. Gone were the rises and declivities covered with wind-tossed prairie grasses. Gone the gentle golden curves of the grass-covered earth. Gone was the brilliant prairie sky, replaced by the sharp cant of cheap wood-frame houses.

And gone too was the woman I had seen from the window of the bus.

War Among the Saguaros

Bruce D. Eilerts

Eventually I became supervisor, responsible for natural and cultural resources management at the Barry M. Goldwater Range, the largest aerial combat training facility in the world. Located in southwestern Arizona, the Range literally lies at the heart of the Sonoran Desert.

The Barry M. Goldwater Range was established in 1941 by a Congressional act under a statute known as Public Law 99-606, which allowed the military to "withdraw" for use in military training 2.7 million acres of public land administered by the Bureau of Land Management. The Range's boundary encompasses most of southwestern Arizona, spreading from the city of Yuma on the border between Arizona and California, east along the Mexican border to the town of Ajo, and north to Interstate 8. After the Range was established, the military's use of the land was still contingent upon compliance with the environmental laws, mandated management projects, and implementation schedules contained in the language of the Endangered Species Act. Here is where I came in, as the supervisor of natural resources management, charged with balancing the use of weapons of mass destruction with natural and cultural resources management. Such a task might be equated to Bambi meeting Godzilla, but despite the many challenges, the military mission continued unhindered because our organization applied the best available science and management practices.

When I started as supervisor there, less than 10 percent of the Range's surface area contained roads, military targets and training facilities. I found the remaining area pristine desert, containing twelve thousand years of archaeological history, a huge intact sand-dune ecosystem, and a variety of wildlife, endangered species, and spectacular geologic features. The land within the Range remained relatively untouched by modern humans due to fifty years of military presence and security. This security resulted in the exclusion of environmentally detrimental land uses such as off-road-vehicle abuse, cattle grazing, mining, development, road construction, bottle-throwing bubbas, and other forms of degradation that are so prevalent elsewhere in Arizona. The Range's archaeological features and artifacts remained intact and unmolested to such an extent that a conquistador lance, intact pottery, and ancient Native American villages were still being discovered there during the mid-1990s.

My experiences with the biologists and archaeologists that I supervised could best be characterized by any number of Gary Larson's "Far Side" cartoons. Although we were a very dedicated and skilled group of professionals, there was never a shortage of eccentricity. Some colleagues even joked that our group was certifiably insane. At least I assumed they were joking. Despite the perceived peculiarity, we became one of the most successful, effective, and awarded natural and cultural resources ensembles within the Department of Defense. Although we were employed by what many considered to be the "evil empire," we were trusted and respected by our peers and most members of the public.

One of the Range's many endangered species that we were charged to protect was the critically endangered Sonoran pronghorn. Less than 120 of these magnificent animals still roam the Earth. All of the surviving Sonoran pronghorn within the United States occur within the Goldwater Range, although occasionally a few of these animals wander onto a neighboring wildlife refuge and national monument. The remainder of the population survives in scattered groups in northern Sonora, Mexico. Unfortunately, these animals are isolated from the U.S. population by roads, fences, development, and other barriers resulting from illegal border crossings and law enforcement.

Consideration of the Sonoran pronghorn is just one of the many responsibilities the Air Force has in addition to training for national defense, since the Air Force also must assume the immense legal responsibility for managing natural and cultural resources within its possession. In one instance, a prominent environmental group filed a lawsuit against the Air Force citing Endangered Species Act violations. The suit concerned protection of the Sonoran pronghorn, but was dropped when it was discovered that the management practices implemented by our wildlife biologists were sound and actually beneficial.

Then in 1996, to the surprise and dismay of many, the unexpected and unfortunate happened. It was a quiet, early morning and the alarm call of a cactus wren floated on the desert breeze. A dust devil danced amongst the creosote and the buzzing of a hovering dung fly hinted that trouble was approaching. An evil presence was about to impose itself upon the land. It was a new Air Force colonel, arriving with ill intent and the sun at his back, much like the Imperial Japanese Naval Air Force had done at Pearl Harbor. The constitutional and environmental antichrist had appeared and was walking amongst us. The colonel had been reassigned to our Arizona air base after serving several years in England. By military standards, such a sudden "backward" re-assignment of a former deputy base commander to a new post while awaiting orders is equivalent to banishment in Siberia. It was clear that the brass had put him out to pasture where they figured he couldn't cause any more harm or embarrassment.

Within days, the colonel viewed his new surroundings and saw a community teaming with lowly civilians in need of military control and discipline. He especially scrutinized the resident group of granola-crunching, posy-sniffing tree-huggers that worked on the Goldwater Range.

Somehow, the colonel had talked the base's wing commander into allowing him to create a new and independent organization called the Range Management Office. He had found his new job, and he was now responsible for overseeing all range-related activities including environmental and cultural resources management. In the military, a colonel, or any other officer for that matter, need not possess professional qualifications or experience to function as an authority or a scientific decision maker. They are experts because they say so! Thus my civilian environmental staff, then attached to the Civil Engineering Squadron, was conscripted into the colonel's new Reich for new supervision and direction. Although all civilian federal employees are governed by the guidelines and policies administered by the federal Office of Personnel Management, the colonel subjected his newly acquired civilians to his own brand of supervisory rules and discipline. His personnel actions and doctrine were often contrary to federal law, not to mention the U.S. Constitution. Civil rights and environmental laws be damned, he made it clear to all of us that the military mission took precedence over all else.

The next three years could only be described as medieval. The colonel eventually engaged in all-out war against my staff and the environment we were mandated to manage and protect. Throughout this ordeal we continued to do our jobs well. We regularly mused over the irony that the Air Force depended on our presence and expertise to prevent its mission from being impacted by environmental constraints. One of the colonel's more notable actions involved his cancellation of established procedures stipulated in a U.S. Fish and Wildlife Service Biological Opinion. This legal agreement allowed military training only if the document's specified terms and measures were implemented to protect the Sonoran pronghorn.

Senior project biologists were replaced with commercial contractors and inexperienced government employees. Accidental bombing and strafing incidents, training accidents, illegal road construction, and unauthorized ground operations within pronghorn habitat were purposely unreported and denied publicly. The truth was regularly misrepresented and records disappeared or were destroyed. Accidental rocket launches, fuel tank jettisons, and aircraft crashes that had occurred on the nearby Tohono O'odham Indian reservation were also handled in a less than honest manner. In addition, the Air Force accidentally dropped live and inert bombs on archaeological sites, environmentally sensitive areas, and several golf courses. One inert bomb landed in the bed of a parked pick-up truck in the town of El Mirage, located more than seventy-five miles away from the Range boundary.

Another memorable incident involved two biologists studying bats in the Range's remote Sauceda Mountains. They were nearly killed when two F-16s on a night mission attacked their camping trailer with 20mm cannon fire. The biologists were six miles from the nearest live-fire targets and had been missed by only a few meters. Such incidents continued to go unreported presumably to avoid liability, embarrassment, and scrutiny. Rampant disregard for environmental laws such as the Endangered Species Act, the National Environmental Policy Act, the National Historic Preservation

Act, and the Sikes Act, the law mandating fish and wildlife management on military lands, had become routine.

Throughout this ordeal my staff and I refused to participate in such shenanigans or to compromise our professional integrity. Letters of complaint to the wing commander, media exposés, Social Actions Office intervention, Freedom of Information Act requests, Congressional inquiries, Inspector General investigations, and box-loads of accumulated evidence did nothing to correct the problem. As our professional and ethical defiance continued, the colonel implemented a campaign of harassment and retaliation. Civil rights violations abounded in addition to the blatant incompetence, waste, fraud, and abuse. Racism, censorship, and religious and sexual discrimination had been documented and reported to no avail. My staff and I continued undaunted for three more years of hell, but the harassment and intimidation were taking their toll on all of us. Concerned about the mounting disclosures of their misdeeds and the resulting threat to their careers, the colonel and his goons singled me out as the ringleader, fabricated egregious offenses credited to my person, and fired me on November 24, 1999.

I became a physical and emotional wreck, and I had difficulty pursuing employment for the next four months. The Air Force even convinced the unemployment office that my purported transgressions warranted denial of my benefits. My entire life seemed to fall out from under me. Miraculously, a Washington D.C. legal-defense organization, advocating the protection and support of civil servants, came to my rescue. Public Employees for Environmental Responsibility (PEER) offered to represent me against the Air Force. With PEER at my side, I slowly began to recover emotionally and I was able to find work as an environmental consultant in Hawaii and Arizona.

Two long years later, in 2001, I won a settlement against the Air Force. I will forever appreciate what PEER did for me. It may sound cliché, but I am certain that the attorneys and staff at PEER literally saved my life. Although the terms of the settlement prohibit me from revealing much about it, I can say that the Air Force amended my official federal employment record to artfully read, "Mr. Eilerts left federal service in 1999 to enter the private economy."

Although the battle had been won, the war seemed to have been lost. The problem Air Force officers were eventually re-assigned or forced into early retirement. To date, the Sonoran pronghorn population continues to decline, and contrary to the restrictions of a Biological Opinion, a road leading into fawning habitat was completely paved during 2002. Off-road-vehicle use is now encouraged along with increased public access to formerly pristine areas. Habitats and archaeological sites are being degraded, and poaching and pot hunting have increased. Uncontrolled Border Patrol and U.S. Customs activities scar the landscape, and illegal immigrants start fires and pollute the few remaining water holes. The Range's remaining natural and cultural resources personnel have been marginalized and the demoralization continues.

The Air Force now spins a different version of the story about my case and the circumstances surrounding it. The Defense Department's twisted account of what happened contends that a disgruntled civilian employee, with an overzealous personal environmental agenda, acted against the best interest of the military. In truth, the case was really about the harassment and illegal, retaliatory firing of a whistleblower who disclosed violations of federal laws, gross incompetence, and rampant waste, fraud, and abuse. It's not hard to understand why the Air Force would rather blame its civilian "environmentalists" as the problem rather than disclose that one of its officers got out of control. However, I will always be amazed by the fact that had the military done something about their problem child colonel early on, they would have avoided years of scrutiny, humiliation, waste of taxpayer dollars, and a damaged reputation.

I admire and support the majority of our country's military. I know that we wouldn't be living in a country of wildlife refuges, wilderness, and national parks without the sacrifices made by the members of our armed forces. My mother watched the attack on Pearl Harbor from her rooftop, and she was fortunate that the United States won the Battle of Midway, which was fought just northwest of her home in the Hawaiian Islands. If it weren't for those heroic battles in the Pacific fought by those few seamen and soldiers who bought our country time to successfully defend it, I might never have been born. It is a shame that there are individuals in our military whose actions defame the armed forces and unnecessarily harm the valuable natural resources of the United States. They do great disservice to us all.

Considering the recent frenzy of political and special-interest attacks on our country's environment, it would be a tragedy if the military succeeds in its current campaign to convince Congress to exempt the armed forces from compliance with environmental laws and procedures because some argue that "readiness" is being compromised. I know from experience that our twenty-five million acres of wild lands within military boundaries can be successfully managed and protected in accordance with law, while ensuring that the mission of our armed forces continues. I can only hope this internal war within our country ceases soon. Otherwise, the casualties will be our dwindling wild places and natural resources. Every one of us will be affected by the loss.

The Troubling Dawn

Estar Holmes

Ford, Washington, is a tiny speck on the Stevens County map, about thirty-five miles northwest of Spokane. It hosts a store with two gas pumps, a post office, and an entrance to the Spokane Indian Reservation. State Highway 231 winds through picturesque Walker's Prairie, past a stone slab that marks the spot where the religious Walkers once settled to teach the local natives a thing or two. Dusty dirt roads lead away from the highway to sparsely spaced homesteads where people still create simple places to survive among bull pines, camas, and deer.

I was enticed to settle on Walker's Prairie in 1982 when I fell in love with a man who had a cabin six miles north of Ford on land that bordered Tshimikan Creek. The first time I went there, he took me to the creek and pointed to the opposite bank. "That's the Spokane Indian Reservation over there," he said. "It reaches to the high water mark on this side."

Tshimikan Creek is a historical gathering place of indigenous people. It meanders through the forested prairie, eventually meets up with the Spokane River, and discharges into the Columbia River. In lovely, secluded places along the creek we swam, and we picked mint for our tea and watercress to eat.

We grew a garden near the cabin, rich with food and screaming with color. We drew our water from the ground and ate the bounty of the land. We conceived and birthed a beautiful son there; it was as if he sprang right from the prairie soil. Those who live scattered loosely among the gentle woods on both sides of the creek know the comforting crackle of a warming fire, the laugh of coyotes in snow-filled winter nights, the chatter of chipmunks, and the rhythm of woodpeckers pounding nearby trees.

Here is a water-hauling, wood-chopping subculture that exists on the fringes of a fading American dream. Refugees from the system come to find some freedom, and it is still possible to live here under the comforting illusion of having done so. The arrangement often entails living in voluntary poverty. Access to water is a struggle. Some people haul water because their wells go dry; others can't afford the luxury of drilling right away.

We hauled water too, and I was gladly living the simple life, working at a nearby restaurant, when the comment of a customer pierced my innocence. I was saying how much fun it is to swim in the creek. His reply shocked me: "Oh, you mean down from the uranium mill that's leaking radioactive water into the creek?"

That was how I heard about Dawn Mining Company and its defunct uranium mill. I had driven by it many times, but it was out of sight just behind the pines. The mill remained nebulous to me, as invisible as the radon that wafts from its tailing piles. I didn't know anything about uranium mills, and I preferred to keep it that way, to focus instead on the pleasantries of day-to-day country life. It was easy to forget about Dawn and to pretend that events unfolding there were not widely important. But thirteen years later, after moving from the prairie, the Dawn cause fell into my lap again. I felt compelled to learn and write about the site and consequently to enter the fight to advocate for its proper reclamation.

I sifted through old newspaper articles, perused environmental-impact statements, and interviewed key players including community members, Spokane Tribal members, and a vice-president of Newmont Mining Company, Dawn's parent corporation. I discovered that Newmont, now the planet's second-largest mining corporation, formed the Dawn subsidiary in 1957 to capitalize on uranium discovered on Lookout Mountain on the Spokane Indian Reservation. Dawn unearthed uranium ore at the mine and hauled it off the mountain to the mill near Ford, on the banks of Tshimikan Creek. For years Dawn supplied the burgeoning commercial nuclear power industry with yellowcake—the moist concentration of uranium oxide that results when raw uranium ore is milled.

In the early 1980s the bottom fell out of the uranium market. Dawn halted commercial production. That's when the long battle over reclamation of Dawn's properties began. The uranium operation had resulted in radioactive pollution on Walker's Prairie. A seventy-acre tailings pit needed to be filled up, buildings needed to be removed, and the entire place needed to be covered with a layer of clay and native plants and monitored for thousands of years. Cleanup of uranium mill sites is expensive, and the bond the company had posted was woefully inadequate. Newmont disclaimed responsibility for funding cleanup of its subsidiary and turned its attention instead to promoting the Dawn site as a national dumping ground for radioactive waste.

The first proposal involved importing barrels of radium-tainted dirt from New Jersey—radium once used to illuminate watch dials. People who lived nearby opposed the plan vociferously. Some of them even offered to donate dirt from their lands to help the strapped company fill its hole. The company rebuffed those offers. When the "not in my back yard" group asked established environmental organizations for help, they were denied on the grounds that the issue lacked "broad significance."

I decided to go see the seventy-acre pit for myself. A dusty dirt road took me to the southern end of Dawn's property, where a barbed wire fence guarded the pit and evenly spaced small metal signs displayed the international symbol for radioactivity. The huge hole, lined with black plastic, contained a toxic stew of 140 million gallons of radioactive water laced with the heavy metals used to coax uranium from its ore. The liquid looked much like any other water, and I mused that deer and birds didn't know what the signs meant. How many people had augmented their diets with animals that had watered at this hole?

After about ten years of state-mandated studies, community meetings, and the deliberations of a technical advisory committee, the Washington State Department of Health denied Dawn a license to bury imported radioactive waste and mandated that the dirty pit be filled with clean dirt. The Health Department instructed Dawn to submit a plan to explain how it would do so, but company leaders refused and continued to pressure the state for permission to dump radioactive waste there.

A proposal was floated to acquire radioactive dirt from various federal sites throughout the nation, as far away as Tonawanda, New York. The final transport plan involved shipment across the country by rail to Spokane, Washington, where a convoy of trucks would complete the trip to Ford—a truck every twelve minutes for at least five years.

About that time I had a conversation with a tribal chair of the Spokane Indian Reservation. "There are a lot of defunct uranium mills like this one between here and New York," he pointed out. "What will stop others from demanding the right to become radioactive waste sites as well?" A couple of years later I had the opportunity to question the head of Washington's health department. "There are no other sites with disposal proposals," he assured me. "This is a completely unique situation."

However, research by the grassroots group Dawn Watch, which formed to monitor Dawn's closure, discovered similar schemes in Colorado, New Mexico, and Utah. Moreover, the mining industry has been pressuring the Nuclear Regulatory Commission for years to relax strict regulations governing what can be buried at old uranium sites. Evidently resolution of the Dawn issue did have "broad significance" after all. If Washington State were to sanction a dumping proposal, other bankrupt uranium mill sites around the nation would be apt to lobby to become radioactive dumps as well.

Meanwhile, testing at the site revealed a plume of radioactive toxins in the Walker's Prairie aquifer. Upon orders of the Washington State Department of Ecology, Dawn pumped millions of gallons of water into shallow plastic-lined pools spread over one hundred acres of the site. The 140 million gallons of toxic stew was added to that, and everyone began waiting for it all to evaporate. As I hauled clean water to my cabin in five-gallon buckets, I appreciated the volume of a million gallons of water and the precious resource it is.

To stop the influx of radioactive waste, Dawn Watch and the Spokane Tribe mounted several legal appeals, all of which failed. Dawn's opponents then pressed the health department to pursue an arrangement the company had originally promised, to make several million dollars' worth of road-safety improvements before the one hundred thousand radioactive-waste-hauling trucks appeared on the narrow two-lane road to Dawn. More studies and public meetings resulted, one of them held on the reservation at Wellpinit, at the base of Lookout Mountain. Company officials and attorneys heard the testimony, their stony faces barely showing a glint of embarrassment, as tribal members once again spoke against the plan.

Spokane native Merle Andrew summed up the tribal sentiment. He and eleven brothers were raised near the Dawn mill, and his family long had lived a little way down the creek from it:

> The whole idea of adding more waste to that piece of land is the most outrageous proposal I have ever heard. I don't understand how the big corporations with all their billions and billions of dollars have to come to the last resting place of my ancestors to drop their waste. It's in line with everything the white man has done to this point. It shouldn't be a surprise to me. But I had to stand up here today and go on the record for the sake of my children and my children's children. If they say that they've had consultation and that none of the Spokane people cared, I'll go on record to say that I was here to voice my concerns. . . .

No radioactive waste has been imported to Washington on Dawn Mining Company's behalf—yet. The pit remains unfilled, the toxic waters evaporating. Pretty soon the people who fought round one will be dying off, but the radioactive legacy will remain for thousands of years. The corporations, which remain legal "persons" into perpetuity, will survive us. My hope is that humans who inhabit this place will always know what really happened here.

Clearwater Journal

William Johnson

For twenty years the author has lived near the confluence of the Clearwater and Snake Rivers in north central Idaho. The following are excerpts from a journal on life in Clearwater country.

September, 1984

Near Bungalow, on the bridge over the North Fork of the Clearwater River, I stand with friends admiring the fall colors. On slopes of the canyon maples have begun to redden, and patches of larch and aspen flare like torches. In the afternoon sunlight the river flows bluegreen, in places almost emerald, over a bottom of sand and stones. Below us, the water is spangled with redfish light, where hundreds of salmon called kokanee have come to spawn. The chemistry of sexual maturity, in which reserves of fat and pigment are depleted for the making of eggs and sperm, has turned their bodies crimson, as if the river holds hundreds of flaming arrows.

Earlier today, five of us—two soil conservation workers, a fish biologist, a high-school biology teacher, and I—rode in a jeep over washboard gravel down French Creek Canyon, toward the North Fork. As members of the Idaho Conservation League, we went to inspect a logging operation on a stretch of Orogrande Creek, a tributary of the North Fork.

Nearly a mile of north-facing slope had been clear-cut to the ridge line. The operation was in its final phase. Skid roads gouged angular scars in the slope, and heaps of slash lay staggered along the road. We passed a diesel Cat parked in the middle of the stream, then stopped for a closer look. Bulldozers and trucks were parked along the road, and raw logs lay scattered in the stream. Sawyers had left no buffer strip of trees, only a grim memorial of stumps where large cedars had stood. The damage, though it occurred on private land, made a case for wilderness protection, for public land would be logged by this and other similar companies.

Even though I'm new to the work of conservation, my more experienced companions confirm what intuition tells me. The land has been badly damaged. The work appears to have been hastily performed, without regard to the future health of the soil or protection of the stream.

The kokanee in the North Fork bear witness to the power and fragility of life and the need for returning home. Some of them would head up Orogrande Creek and discover that home was not as it had been before. On the drive back to town, images clash in my mind. Slivers of redfish light tell me there is beauty and hope in the river. The stumps and slash on Orogrande Creek remind me that I can never underestimate our power to destroy.

October, 1985

I'm a guest at a Potlatch Community Advisory Council luncheon. Potlatch Corporation oversaw the logging on Orogrande Creek, on land it owns. Today, corporate representatives will respond to our complaints and questions. Our complaints are threefold:

1. The extent of the clear-cut (nearly a mile in length) will increase siltation and so harm spawning habitat.
2. The slopes are steep—skid roads will likely erode.
3. No buffer strip (specified in National Environmental Policy Act regulations) has been left to border the creek, whose water temperature during summer will rise, with potential harm to fish.

But these formal, observable facts represent the tip of an iceberg, and as I listen to the hydrologist make his predictable case, I hear a subtext of my own.

The Potlatch hydrologist is well prepared. His slides present the corporation's view. The slopes were clear-cut *because* they were steep, because leaving a few trees standing would have made them susceptible to blowdowns. Most of the trees cut were already dead or dying—it was either cut now or risk greater damage by fire. Trees along the stream were cut because it was feared they'd catch fire (again because of the steep slope) when the area was burned to nourish the soil for replanting. A slick pamphlet entitled "The Orogrande Conversion" explains all this and adds that the company will replant the slopes and work to improve fish habitat in the stream.

I respect the work of loggers when it is responsibly performed, and I have walked through forests once logged that retain their diversity and character after a period of healing. But on Orogrande Creek, the wounds are too deep, quite possibly terminal. What Aldo Leopold once called the "integrity, stability, and beauty of the biotic community" has been irreparably harmed.

I look at the faces in the room. We are all trying to be patient, trying to listen, even to hear. But there is a gulf, a great divide between those wishing to protect the land and those who see it as a material resource. We might as well speak distinct dialects of English, and perhaps we do. The assumptions we hold about land, forest, and management, even the words we use—"forest," "use," "profit," "wilderness"—seem mutually unintelligible. Where, I wonder, is the middle ground, the domain of fruitful compromise? If a corporation can defend land use like that on Orogrande Creek, it can defend anything. I pale thinking how the common citizen might counteract such arguments, let alone the money and power behind them.

October, 1986

I see the forest as an extension of, and a presence in, my life. To me, a forest is a living process, one that cleanses the air I breathe, protects and purifies the water I drink, and lifts my spirit and frame of mind. When such a view, and it is not uncommon, informs the lives of those who manage the land, the land will be managed responsibly.

December, 1987

Too often the corporate view conditions people to single-vision, a monolithic way of seeing. It is the way of conventional technology, a rationalized standoff with nature that aims to control. The desire for profit blurs the line between use and abuse. Profit doesn't require abuse per se, but if profit is at stake abuse is justified. But if we understand the forest as a living environment, an extension of ourselves and we of it, then we're talking about self-abuse.

When the kokanee spawn and die in the North Fork their bodies will replenish the stream with nutrients—food for bacteria, insects, other fish, birds, and small mammals. The same is true of dead wood (the cultural sense of the idiom is telling), which nourishes the soil, from which new forest springs. The dead give of themselves so that others may live—these ecological relationships have spiritual significance in a cycle of death, birth, and resurrection. But it depends on how we see.

✦ ✦ ✦

Orogrande Retrospective

July, 1983

In the shadow of cedars I sit on a log, tossing pebbles into Orogrande Creek. Small trout dart through a clear pool over stones and clean white sand. The creek is mottled in light and shadow. The day is warm, but shade and flowing water cool me. It is good to be alive in such a place, and I am thankful it is here. So, I think, are the bright yellow butterflies that wander about in scrubwillow near the bank.

July, 1984

The Orogrande project occurs.

Summer, 1985

The logged slopes along Orogrande Creek are burned and replanted with grand fir.

July, 2001

Some of the replanted trees are now between five and eight feet high, forty years or more from being harvestable as timber. This stretch of the canyon remains dry and dusty, a place I want to hurry through, which has the look of waste and wear. Common sense and a pair of eyes say it holds considerably less biological diversity than it did before "the Orogrande Conversion." Its stability, integrity, and beauty are not apparent—quite the opposite.

✦ ✦ ✦

Undated Entry

A popular bumper sticker reads *Wilderness—Land of No Use*. Idaho contains some four million acres of designated wilderness and could easily contain more. The assumption is that if you can't road or log it, it's of no use. But the word use too easily begs the question here. We can't, and we shouldn't, measure it by profit alone. We breathe, drink, hunt, fish, berry-pick, hike, and muse to use this world. As the poet W. S. Merwin puts it: "we were not born to survive / only to live."

June, 1988

The corporate problem is one of scale. The larger the entity, the more abstract; the more abstract, the less likely to cherish and protect the small, the intimate, and the local.

A forest is not an idea. It is a living biological reality—intricate, complex, interdependent; it includes rocks, logs, trees, roots, water, fish, beaver, kingfishers, on and on, into the dense and subtle fiber of creation. Such reality, or presence, can't be translated into an abstraction without grave risk, but this is what happens when people who lack intimate knowledge of the land make economic decisions about it. The Orogrande's "argument" lives in our blood; it bubbles in the Paleolithic strata of our genes. The *eco-* in both "ecology" and "economy" once meant *home*.

SECTION III
VICTORIES

Lost and Found

Terrell Dixon

It was the beginning of spring in the early 1990s. At ninety-five hundred feet in Rocky Mountain National Park in northern Colorado, Linda and I looked out over a nine-acre alpine lake. The snow line was about eighty-eight hundred feet, and we had struggled up the last section of trail, postholing through snow that was thigh-high.

Except for the jays and squirrels announcing our arrival, the area was quiet, an increasingly rare tourist-free time in a 415-square-mile park where the annual visitor count was pushing toward the three million mark. Later in the spring, the trail to Fern Lake would be a lush green, following rivers and streams much of the way, passing two large waterfalls and a variety of ferns, which give the lake its name. There would be mountain chickadees, black-capped chickadees, and blue grouse hens clucking softly to guide their chicks. On that day, though, more than three-quarters of the lake surface lay frozen, and the sight of that familiar place in a new jewel-like guise proved dazzling. The ice stretched smooth and silver across the lake, broken with patches of emerald-green water, both ice and water set close against the gray heights of nearby Notchtop Mountain and the Little Matterhorn. A new fly-fishing season began for me that day. It had been a long winter's wait, and I quickly started to set up. Thick gloves made accustomed acts seem new; simple things—attaching reel to rod, threading the line, tying on the fly—seemed to take forever.

After all that, the fishing was disappointing. I knew that the fish were there. At the end of the previous summer, Linda and I had hiked up to the lake to watch the fish. There was a unique kind of peace in seeing them fin their way through the shallows, perfectly at home in their watery world. On this day, however, the fish were keeping to themselves, so we walked around the lake, ate a sandwich, and soaked up the beauty of the place until thunderclouds and cold told us it was time to leave. Packing up, I discovered—

too late to do anything but make a cursory search—that the tip section of my fly rod, a thin piece of graphite about sixteen inches long, was gone.

So the next morning we hiked again to Fern Lake to look for the lost rod tip. This resolve to search for a slender strip of graphite, in that snowy expanse made even larger with the previous day's fresh snow, seemed odd to our friends—an overly optimistic and even quixotic effort. For us, however, it fit the pattern that had emerged in our lives over the past decade. In the eighties, we had been fortunate to travel parts of "the big wild" of Alaska. We floated the Noatak, the Alatna, the Kongakuk, and the Jago rivers, and we hiked and backpacked the landscapes of the Arctic National Refuge through which they flowed. We saw musk oxen gather into a protective circle and gaze out at us through the fog and mist of the Arctic plain, watched a wolf loping after an isolated yearling caribou, enjoyed the sight of grizzlies at a distance, and survived two close grizzly encounters. What was once a distant and fairly abstract wilderness became a place we knew and loved.

It was also, then as now, a threatened landscape. The mixture of resolve and hope that was then our way of dealing with the drilling crisis took shape after a backpacking trip to the Arrigetch peaks. On that trip, crucial pieces of equipment—a stove valve, a hiking boot, a pair of binoculars—were given up as lost, sometimes for as long as a week, and then, incredibly in that vast expanse of wilderness, found. These experiences fed into our concern for that remote wilderness area and shaped a personal rallying cry, one that we wanted to prove as true for the land as it had for our gear: "Nothing is ever lost in the Arctic."

The early springtime attempt to bring the personal, practical part of this resolve to Fern Lake in the Rockies paid off. Although the snow on the second day was deeper, our tracks were not entirely covered, and I finally located the rod tip under the sheltering branch of a Douglas fir, dusted with snow, looking remarkably like a slender tree branch. Since the sun was out at mid-day, I celebrated by taking a few minutes to fish before we went down.

On this second day, there were fish to be seen moving about in the green patches of open water and sometimes visible also under the thinnest portions of ice. Now, nearly a decade later, I remember what happened next with an intensity that matches that of the earlier encounters with the wildlife of the Arctic National Wildlife Refuge. I could see, some twenty feet away in an open space, a trout finning its way around the lake, and so I sent a bushy fly out to another patch of open water directly in front of me. In the clear water, I could see a deep green back with large black spots and a bright vermilion on its lower sides. It moved toward my fly. After an excruciating wait, the fish's easy movement ended and it shot to the surface to take the fly.

In the next few moments, as I slipped my hand into the icy water to release the fish, my sense of the world changed. This was not due to the size of the trout. It was a healthy fish, but a small one, perhaps nine or ten inches long. Nor did its impact stem from its beauty, although it was a beautiful fish, exceptional even in the world of mountain water where the dappled glories of trout are widespread. No, the change came from an even more basic element: the presence of that fish. It is almost a commonplace among those of us who meditate on the meaning of fly-fishing to observe that one of its defining joys is the tangible, vibrant connection to the wild that comes with the tug of life from the other end of the line. This fish, however, gave an added dimension to that familiar feeling. It was a Colorado greenback cutthroat trout, a species once native to this part of northern Colorado but that had for decades been given up as lost. Its presence in Fern Lake that day signaled an enormous, beneficent change in our human relationship to that landscape.

Before the time of European settlement, the greenback cutthroat trout lived east of the Continental Divide in the watersheds of the Arkansas and Platte rivers. With settlement and ever increasing development, their range and numbers declined drastically. The reasons are unsurprising: water was taken for other uses, streams were polluted, and settlers, choosing to believe the Western myths of unlimited abundance, fished the species to the extreme edge of extinction. Some early efforts at hatchery restoration were made in the last decade of the nineteenth century. When these efforts failed, other non-native fish more easily adapted to hatchery stocking programs, rainbow trout and Yellowstone cutthroat trout, were stocked in huge quantities. They began to hybridize with and displace the native greenbacks. By the 1930s, some thought the greenbacks were extinct.

This long decline and assumed demise began to change some thirty years ago. A very small population of the fish turned up in 1968. The Endangered Species Act passed in 1973, and the greenback was listed on it. Extensive surveys throughout Rocky Mountain National Park found a few more isolated pockets of greenbacks. Then, with less than two thousand total fish inhabiting only about three miles of stream habitat, a recovery program began. Using these remnants and improved hatchery techniques—along with the sustained cooperative efforts of the National Park Service, the Colorado Division of the U.S. Fish and Wildlife, and volunteers from Trout Unlimited—the Greenback Recovery Program gradually became a success. Initial public resistance gave way to pride in a resilient and beautiful native fish, and in 1994 the Colorado greenback cutthroat trout became the official state fish of Colorado. It now occupies some thirty-three streams and lakes in Rocky Mountain National Park, and many of these populations are self-sustaining.

Threats continue, however. The general problems of global warming, nitrogen deposition, and whirling disease are compounded by accelerating development along Colorado's Front Range and by the related increase of tourism in the park. The peaks and lakes of the park retain their striking beauty, but in the summer and on weekends the elk and moose jams along the roadside can resemble city congestion. In the distorted logic of government spending, this overuse is accompanied by underfunding; many necessary research and repair projects remain undone. In the midst of steady assaults on national parks, the return of greenback trout holds hope.

In the decade since our time in Alaska, it has become clear that some things truly will be lost, despite our desire to be optimistic. We live in a time when habitat loss is endemic and when species around the world are disappearing irrevocably at an alarming rate, and we are, once again, incredibly, in danger of losing the Arctic National Wildlife Refuge. That Colorado greenback cutthroat trout at Fern Lake, however, taught me that the sometimes-gritty labor of restoration can work its own kind of magic. We must preserve the wilderness riches of the Refuge, but we must also attend to the work of restoration in the lower forty-eight states. Wilderness is often viewed with a special reverence, and rightly so, but these times require the realization that there is also grace in the *restored* wild, that dedication to recovery and restoration must accompany preservation.

Great Sand Dunes: The Shape of the Wind

Stephen Trimble

Jake leaped through the mirage at the lip of the dune—and disappeared. Had he passed into another dimension?

I followed, to see my eight-year-old son plunging down Great Sand Dunes, running, jumping, flying over this Colorado surprise. He had never been to this island of sand that crests at 750 feet above the floor of the San Luis Valley, the tallest dunes on the continent. But he knew what to do.

The delight of plummeting down a slip face so thrilled him, he forgot to complain about the ache and stretch when we had to climb back up; he almost didn't mind the wind-whipped sand that stung our bare calves when we reached the crest of the high dune.

Our senses reassert themselves in this place. In the nearly complete silence, we hear the wind. We hear other sounds of wildness: a hermit thrush, flutelike and invisible in creekside woods, the soft pads of cottontail feet lightly drumming the earth, the drunken buzz of a bumblebee wallowing in thistle pollen. We feel with a shock the cold of Medano Creek as we wade ankle-deep on our walk to the dunes. Darkness, real darkness, fills the night sky with stars, the garland of the Milky Way unreeling, bringing us closer to other worlds.

Great Sand Dunes National Park and Preserve balances pairs of opposites. Aridity and moisture, heat and cold, light and darkness, curve and angle, change and constancy, desert and mountain, windstorm and stillness. Alpine lakes rest serene in wildflowered meadows, while below lie prairie and desert, cactus and tarantula.

Here, high, lonesome, treeless basins—called mountain parks—span vast depressions between Colorado mountain ranges. Local boosters describe this park, the San Luis Valley, as "the highest largest mountain desert valley on the North American continent." They are right.

Reaching into New Mexico from south central Colorado, covering an area nearly three times the size of Delaware, the valley floor sweeps away in smooth planes. These uncompromising flats angle sharply upward at the valley's margins in the fourteen-thousand-foot peaks of the Sangre de Cristo Range and the San Juan Mountains, piercing an achingly blue sky.

One large light-colored area at the valley's eastern edge contrasts in tone and texture with everything in sight. Its colors shift with the light—cream, gold, gray, tan, even pink. Here, six thousand vertical feet of rock and forest and snow meet the soft lines and uninterrupted colors of sand dunes. This wild variation in elevation as the Sangres rise steeply from the valley helps make these severe mountains the range richest in species diversity from Montana to New Mexico. It also compresses remarkable recreational diversity into a narrow corridor between dune and mountainside.

From the trickle of meltwater in the alpine snowfields of Mount Herard, Medano Creek plunges through Sangre de Cristo forest and into the valley sun, past the blue intensity of Colorado columbines to the arid miracle of creamy blossoms rising from spiky yucca. Along the edge of the dunes, the streamside corridor makes the ideal foreground for postcard views as well as the "beach" playground. From the campground or picnic area—reached on good paved roads—splash across the creek and climb into the dunes. Stroll just two hundred yards to a swale laced with kangaroo rat burrows and swaying with delicate clumps of Indian ricegrass.

When you do so, you enter the 87 percent of Great Sand Dunes designated by Congress as wilderness in 1976. As Kathy Brown, chief of interpretation at the park, says, "People who are not traditional back-country users cross that stream and they are in the wilderness. Folks who might not otherwise have the chance can have a hands-on wilderness experience. With surprising diversity, this is a wonderful place to explore."

Medano Creek is a knife. Its waters keep the dunes away from the mountainside, continually slicing at the leading edge of sand. The creek has a surging flow that fascinates visitors and makes it an unusually powerful sand transport. The creek spreads out along the base of the dunes, meandering in braided stream channels. Water in each channel pushes up small ridges. Each ridge grows higher, holding back more water, until the water breaks over its top in a wave, surging downstream like breaking surf.

When the creek dries up in fall, these new deposits are blown back into the dunefield. Dunes along Medano Creek become more than a simple series of dunes, backdrop to the entrance road. They are a mountain of sand. Today, we know that Great Sand Dunes is virtually a closed system, with sand spiraling and recycling through the creeks, then blowing back to the dunefield—but with little new sand arriving from beyond.

The creek recycles. The sand spirals. The dunes dance. Trapped in the vortex of wind, water, and mountain, dunes form, shift, and grow. Their future depends on water—the groundwater system of the valley together with streams splashing down from the mountains.

The dunes lie at a rich cultural crossroads, as well. The adobe and spice of the Hispanic Southwest reach north, halfway up the San Luis Valley, where they give way to the Anglo Rocky Mountain West.

Within a year after Don Juan de Oñate claimed New Mexico for the Spanish crown in 1598, "from the leaves of the trees in the forest to the stones and sands of the river," the Spanish reached the northern San Luis Valley. On one of these early trips, tradition tells of a Spanish priest—maybe Father Francisco Torres, maybe at San Luis Lakes (now a state park just west of the dunes). He and his companions had enslaved their Pueblo Indian guides at new gold mines; the Pueblos allied with the local Utes and rebelled. As Padre Torres lay mortally wounded in his comrades' arms, he lifted his eyes to the mountains rimming the valley on the east, blood red in the last light of sunset. The devout man whispered, "*¡Sangre de Cristo!*" (Blood of Christ!), and died, giving the mountains their name.

Nineteenth-century settlers viewed the "useless" dunes as a fine refuge for wild horses, which—legend had it—developed webbed feet from their years living on the sand, enabling them to easily escape from ordinary mounts. Other legends of the dunes spice local lore: Spanish oxcarts, forgotten towns, murderers, herds of sheep, lost travelers, and shipments of gold—all buried under shifting sands.

You still can ride horses into the dunes, even if they don't have webbed feet. Our wrangler, Quentin Dickey, led us across the creek and into sunflowered bowls beneath cumulus clouds building toward their afternoon climax. He entertained us with stories from his movie stuntman days, pinpointing the scenes in *McClintock*, *True Grit*, and *Rooster Cogburn*, where John Wayne killed him or punched him or sent him careening down a muddy hill.

In the same spirit of time-gone-by, Great Sand Dunes was originally preserved as a curiosity, "nature's sandbox." One 1924 travel writer called the dunes "the eighth wonder of the world." President Herbert Hoover's 1932 Great Sand Dunes National Monument began to receive a multiplying stream of visitors, reaching three hundred thousand annually in the 1990s.

Today, we are re-imagining the concept of national preserve. "We've gone from curiosity to . . . big ecosystem landscapes," in Secretary of the Interior Bruce Babbitt's words. Grand Canyon-Parashant and Giant Sequoia National Monuments, proclaimed in 2000 in Arizona and California, respectively, extend federal protection to whole ecosystems rather than just the most conspicuous central features. Grand Staircase-Escalante National Monument, created in 1996, embraces six million acres of Utah largely for scientific reasons—to preserve more than one sample of each environment in the huge research area.

At Great Sand Dunes, the same vision has expanded the limits of the monument to its natural boundaries, expanding upward to the ridgeline of the Sangre de Cristo Range and outward into the valley's alkali flats and wetlands.

In the 1980s, a proposal to sell two hundred thousand acres of valley ground water annually to Denver's exploding population threatened the valley's resources and rural lifestyle. Farmers, ranchers, environmentalists, and the National Park Service responded by allying in a landmark New West coalition. When scientists predicted that Medano Creek would not reach the dunes at all with the proposed level of pumping, the water raiders could not proceed.

In the flurry of recent research, scientists have rewritten the story of the sand, grain by grain. Begin with Rio Grande floodplain deposits on the floor of the San Luis Valley. From here, sand blows toward the Sangre de Cristo, funneling into a bay where the range recedes eastward. At the apex of this alcove, the southwesterlies pour over three low passes but cannot drive the sand up the steep mountain slope through dense forest. So the wind drops its load here. Dunes form.

For decades, we believed that these three elements explained Great Sand Dunes: source, wind, trap.

We now know that every Rio Grande sand grain passes downwind into a complex system of sand deposits covering almost 350 square miles—far beyond the thirty-square-mile dunefield itself. The national-monument-sized Great Sand Dunes that existed from 1932 until 2000 protected only one-tenth of this windblown system. Ecologists rank the unusual insects and wetlands of the Great Sand Dunes landscape as "globally significant" in biodiversity.

Action has come quickly. In 1999, the Nature Conservancy bought the one-hundred-thousand-acre Medano & Zapata Bison Ranch adjoining the dunes to the south and closed its golf course. The next year, the Colorado congressional delegation introduced legislation to expand the boundaries of Great Sand Dunes. Their bill, passed and signed into law in the fall of 2000, authorized the citizens of the United States to buy the one-hundred-thousand-acre Baca Ranch to the north of the dunes, protecting the aquifer, wetlands, and wildlife. These new public lands will add high country to Rio Grande National Forest, create a national wildlife refuge from valley wetlands, and enlarge the monument into Great Sand Dunes National Park and Preserve. It is a time of change.

Travelers to Great Sand Dunes enter a community that leads outward from the dunefield in relationships that spread like ripples in San Luis Lake. To the Nature Conservancy's bison ranch, to twelve thousand home sites on the Baca Ranch, to Buddhist retreats and crystal shops in the New-Age village of Crestone, to precious valley aquifers that nourish traditional farms, to 235,000 acres of designated wilderness in the Sangre de Cristo.

Immersed in a living, changing Great Sand Dunes, we remain, always, at the threshold of discovery. On the edge, on the border, at the crossroads of the West. At a magical equilibrium between wind and water and sand.

A Little Garden of Sand:
Bringing Back San Francisco's Native Dunes

Christine Colasurdo

I live on top of a sand dune—a really big sand dune. It begins at the Pacific Ocean and fingers its way seven miles east to the San Francisco Bay. In the nineteenth century people called it the Great Sand Waste. It swept over San Francisco's hills and valleys like snow, blanketing peaks six hundred feet high and accumulating into drifts one hundred feet deep. Its soft curves were furry with silvery lupines, beach strawberry, dune tansy, and lizard-tail. Vernal pools collected in its depressions and reflected back the fog of San Francisco's sky.

Today, most people living in San Francisco aren't even aware that this vast dune system exists. Entombed by development, it lies buried beneath houses, stores, skyscrapers, and streets. Its vernal pools are gone, and most San Franciscans wouldn't be able to name a single plant indigenous to the original dune ecosystem if their lives depended upon it. In fact, many people in my neighborhood try to cover up what remains of the sand dunes in their yards. Some pave their front yards entirely, then paint the pavement green. Others install green plastic turf, then decorate the borders with pots of red plastic roses. Still others cover the ground with black tarp, then install rock gardens, and—*voila*—no trace of dune can be seen.

It's no wonder, then, that San Francisco has lost 355 of its native species since 1776. Some species, like the presidio manzanita, are teetering on the edge of extinction. Others, like the Xerces blue butterfly, are already gone. Last seen in 1943, the Xerces blue became North America's first known butterfly to go extinct. Some animals and plants, like the California quail and the "good herb," *yerba buena* (for which the early village of San Francisco was named), have essentially been extirpated and are best seen as rarities in the city's Strybing Arboretum. Mammals like elk, which once crossed the dunes on their way to good forage areas, are also gone. Even grizzly bears once visited San Francisco's dunes.

We can't bring back grizzlies to San Francisco. But several years ago, it dawned on me that perhaps I could bring back a bit of sand dune in my own small way. As I walked to catch a train to work one morning in 1994, I was struck by the unusual ugliness of a schoolyard two blocks from my house. It wore the neighborhood's trademark aspect of forgotten dune—dry, compressed earth where only wisps of weeds grew. The grounds belonged to Lawton Alternative School, a public school from kindergarten through eighth grade. Each time I passed, I couldn't help staring at the neglected corner of schoolyard. The pale stubble of dead lawn and invasive weeds contrasted pitifully with the handsome Art Deco building, with its luminous glass tiles and copper-edged eaves.

Then it struck me: perhaps I could build a garden there—not just any garden, but a garden of native plants tough enough to survive the rainless summers, chill oceanic winds, and damp summer fog that sweeps over the neighborhood almost daily from the Pacific.

After talking to activists who were already volunteering with children, I decided to get my hands dirty. In September 1996, I submitted a garden proposal to Lawton School. Luck was on my side: the school had already decided to undertake a major gardening effort and had designated someone to manage the task. Within a week of submitting my proposal, I began planning the site of my future garden with Carol Lanigan, the gardening coordinator. The native garden would become one of several projects created that year by the students themselves.

The plot chosen for the garden was a narrow, south-facing strip about forty feet long, with a thirty-degree slope. In December, Carol organized a crew of volunteers to break up the compacted soil and uproot weeds. During the same month, Carol and I went on a quest for plants. The school lacked money, so donations were a priority. I managed to transplant many natives I had cultivated in my back yard—including California poppies, pearly everlasting, California lilac (*ceonothus*), yellow bush lupine, *artemesia*, golden aster, and yarrow. But that still left many other plants to procure.

Of all the challenges confronting us, obtaining plants proved the most difficult. Many Bay Area nurseries do not stock native plants or, if they do, the selection is minimal. On the other hand, it is considered unethical to take seeds or plants from the wild. So we found ourselves horticultural beggars. Fortunately some nearby nurseries, both private and public, donated several plants. By the end of autumn we had pots of beach primrose, blue-eyed grass, Douglas iris, California fuchsia, California lilac, coast silk tassel, coyote brush, red currant, seaside daisy, and sticky monkey flower to add to my backyard transplants.

Two classrooms enrolled in the project—Lauren Skye's lively fourth-graders and John Gough's soft-spoken fifth-graders. With three potted natives and a poster in hand, I visited each class to introduce myself. As I passed around the plants, I explained the concept of gardening with indigenous plants—plants that are adapted to the soil and climate and don't need extra water or fertilizer. I was surprised at how energetically the students grasped the concept. Dozens of small arms shot skyward within seconds of each question I asked: What is a native plant? Why are native plants important? Why are some native plants disappearing? Not every answer was the one I was looking for, but the enthusiasm in each student's face was infectious.

With the earth tilled and winter rains abundant, it was time to plant the garden. On a cold January morning in 1997, the students lined up to "adopt" their plants. Holes were dug with the help of adults, and after much crouching, shouting, and sharing of tools, the plants were in the ground, some a little more cockeyed than others. Although spirits ran high, I was surprised to see how hesitant some children were to get their hands in the earth. I could only conclude that this was their first gardening experience.

The following week, students wrote the name of their adopted plants on wooden stakes and hammered the stakes into the ground next to their plants. Carol organized another volunteer crew to spread mulch throughout the garden to retain moisture and keep weeds at bay. After a few months of limited watering to help the plants get established, the garden was left to its own acclimation.

Since that first year, vandals have broken stakes, some plants have died, and weeds have seeded in. Every school year I have continued working with each new crop of John Gough's fifth-graders, assigning plants in the fall and helping students to yank weeds, pick up litter, and replace stakes. Throughout the year, the students track their plant's growth in journals, where they keep a pressed leaf or maybe a bloomed-out flower. They note changes in their plant's height, width, and color and size of leaves. They also learn about their plant's Latin name and other botanical details. Some excitedly record their plant's prodigious spring growth, making me wonder if they have ever before watched nature so closely. Excited whispers and even applause greet my entrance into the classroom each month—perhaps because I've just rescued them from a difficult math assignment, or because we're going outside.

With the plants blooming during San Francisco's mild winter and dying back during the summer, the garden's seasonal rhythm works out well for the academic year. When the students start school in September, the plants are estivating; the poppies and pearly everlasting have gone to seed, and the red currant and *ceonothus* have produced berries and lost some leaves. Then, as the students adopt their plants, the autumn rains arrive and subtle changes occur: new buds show on some shrubs, and tender blades spring up on the blue-eyed grass and sea pinks. By January and February the garden begins its early spring: the red currants unfurl their pink waxy flowers, and the poppies shake out luxuriant, lacy leaves. While northern gardens are hunkered down in frost and snow, our little bit of dune is exploding with new growth. By late April, everything is in full petal. Then, as the students finish the school year in June, the garden prepares itself for the impending summer drought. In that way, the plants echo the activity of the school year, "going to sleep," as one child put it, while the students are away, and expanding vigorously while the children are nearby in the classroom, expanding their own world through learning.

I, too, expanded my own world through the garden. Several years after starting my small project, I learned of a larger gardening project that the National Park Service had initiated in 1990 only a few miles from Lawton School. Not unlike my small plot, portions of the Golden Gate National

Recreation Area—the largest urban wildlands in the country—were being restored. The National Park Service, along with the Golden Gate National Parks Association, had begun to rip out invasive exotics in federally owned coastal bluffs and other wild areas and restore them with lizard-tail, lupines, dune knotweed, seaside daisy, and other shrubs and wildflowers. At Fort Funston, the Presidio, the Marin Headlands, and Crissy Field—all former military sites—the National Park Service was coordinating a massive volunteer effort with teenage students and adults. In some areas, such as Fort Funston, acres of dunes formerly covered with one exotic species—South African ice plant—now have more than thirty different species of indigenous plants. With the official reopening of Crissy Field in May 2001, the National Park Service has restored more than fifty acres of dunes and wetlands right in San Francisco. At Crissy Field alone, more than three thousand volunteers have uprooted weeds and planted thousands of native wildflowers, shrubs, and trees. And more than twenty-four acres have been transformed from paved, weedy grounds to flower-dotted dunes, uplands, even a salt marsh alive with shore birds.

It has been five years since I started the garden, and I still bump into some of my former students around the neighborhood. They wave hello, and although I can't always remember their names, I ask them if they still remember the name of their plant. They always do. I like to think that maybe in remembering their plant they carry a little of the forgotten dune inside them. If so, may it bloom.

Many thanks to Esta Kornfield, Barbara Pitschel, Pete Holloran, Jake Sigg, Greg Gaar, David Graves, and others who have shared their knowledge of San Francisco's native species and advocated on their behalf.

Postcards from the Pleistocene: Saving Hendrickson Canyon

Robert Michael Pyle

In 1973, as Deep River resident Jack Scharbach could see, logging had ruled in the Willapa for many decades. The watersheds in the Willapa Hills—the low, rainy coastal range in southwest Washington—had been cut over, leaving a tatterdemalion landscape of recent clear-cuts, young plantations, and extensive second growth. But the forest behind Jack's place was different. He wandered freely in those deep woods, getting to know the ridges and ravines intimately. The elk got fed up with his following them, and one day a large bull charged him. He knew it was a special place he'd landed.

Jack had happened on the only stand of original forest in all of Wahkiakum County. The old-growth in Hendrickson Canyon stood out like a pompadour in a company of chrome-domes, so bald were the surrounding sections of private timberlands. The canyon, in contrast, belonged to the people of Washington. The Department of Natural Resources (DNR) administers the state forests on behalf of a number of trusts. Hendrickson came under a "university trust," one designated to produce income for the five state universities. Luckily its stumpage had never been put up for bid.

Washington Natural Heritage Program biologists Reid Schuller and Rex Crawford confirmed Jack's find and recommended Hendrickson for protection. When I learned about this, I rushed to see the canyon. But the maze of logging roads that pierced those shaven acres proved so convoluted and the maps so contradictory that finding the forest turned out to be a real challenge. Several failed forays made it easy for me to imagine how Sasquatch can live undetected in the tangle of successional forest around Mount St. Helens. My search also showed how tiny the leavings really are, that they can be secreted among the immensity of the second growth, despite the size of their great old trees.

Once I had located the place, many field trips followed. Now, when I think of Hendrickson Canyon, I think of which flowers were blooming, what birds were singing, and who saw and heard them with me. During one late-autumn outing, bright chanterelles and even oranger polypore fungi daubed the secret places with unexpected color. Penny-whistling winter wrens shattered the same silence the varied thrush only shivered when it gave its hoarse call, like a whistling through spit. Slanting great cedars hung draperies of moss mixed with their own lacy foliage, a green screen to the pinched winter light that rarely reached the ground through the canopy of conifers. Winter was on its way out when I took my friend and mentor Charles Remington, of Yale's Peabody Museum, to see the trees. We sought the wood roach, a creature merging evolutionary traits of both roaches and termites; we found many-legged myriapods—centipedes and millipedes—attenuated marvels of nervous organization. There were isopods, related to common "pill bugs" or wood lice, their young collected around their legs in a display of parental care uncommon among invertebrates. And there were snail-eating ground beetles, their mandibles elongated to reach into coiled snail shells, and several species of their prey, notably the Vancouver green snail. Charles collected a rich bag of insects. I promised to keep searching for the wood roach.

Spring seems the liveliest time in the forest, and our expeditions center on that season. My wife, Thea, and I took to the canyon to catalog trees one March. Hendrickson is the truly mixed forest of the very old Willapa inland—some spruce, some cedar, a lot of hemlock, and remnants of Douglas fir. The largest tree we found that day, in fact, was a massive Douglas fir. We lounged by the big tree's base and pondered the wealth of the uncut woods, a treasure not measured in board feet but in the number of notes in a winter wren's song and the legacy of leaflets in the carpet of oxalis shamrocks spread before our boots. When we learned this tree stood a little outside of the trust parcel, and it later fell to a logging contractor who also bulldozed the nearby stream, we were shattered.

The next spring outing to Hendrickson found a band of old-growth pilgrims assembling at Swede Park, our home in Gray's River, on an April morning. Elizabeth Rodrick of the Washington Department of Game's

Nongame Wildlife Program had come to see the forest as a possible habitat for northern spotted owls. The party included Carol Carver, Wahkiakum County extension agent; David and Elaine Myers, local photographer and master gardener; old-tree tracker Bob Richards, and others. The hoped-for owls failed to respond, the habitat too small for them to breed. But we found amphibians—red-legged and chorus frogs and several salamanders. A small, spring-fed stream led to a spruce and cedar cleft where a pond stood, its water roiled by massing newts. Caddis fly larvae dragged their shelters across the newts' gelatinous egg masses. We reached a broad ridge carpeted with grass-green oxalis and single-flowered wintergreen, with its heavenly scent, overhung by cedar snags from a fire. Remains of three deer could have been cougar kills, as they were not far from where I had once seen the big chocolate-colored cat. Reluctantly we rounded the ridge and stepped back out to a clear-cut and cars.

The year at Hendrickson came around with a late-autumn excursion in search of spiders. Our arachnologist friend, Rod Crawford, extended Professor Remington's invertebrate list with his own careful survey of spiders and other microfauna. In a brief circuit of one lobe of forest, tangential to the main old-growth stand, Rod found twenty-two species of spiders and six kinds of harvestmen, including one new state record and one old-growth indicator species. All this, on a rainy day, after most of the natural activity had shut down for the season.

Besides enjoying and studying the canyon, we accelerated efforts to protect it. In 1983 we got a resolution, from the Columbia-Pacific Resource Conservation and Development District, that urged DNR to reserve the canyon as an old-growth natural area. In 1984, the Wahkiakum County Democratic Convention passed a similar resolution. In 1985, the Wahkiakum County Board of Commissioners did likewise when Joe Florek (a logger), Bob Torppa, and Kayrene Gilbertsen all voted for the measure. The commissioners went on record as "supporting the preservation of the Hendrickson Canyon Old-Growth Forest to protect the same from being destroyed or severely impacted by the encroachment of man for uses other than educational and conservation purposes." Even so, Crown Zellerbach opposed the Hendrickson designation *sub rosa*, apparently just on principle, or maybe because a preserve plunked down beside their managed land would be inconvenient.

The fact that old-growth indicator spiders occur in the canyon might not impress anyone in the local tavern, but the heritage idea made sense. Dennis Nagasawa, neighbor, fire chief, and local DNR man who kept a protective eye on the stand, told me that cones had been gathered from big trees in Hendrickson for seed recovery. Developed through natural selection over ages, the genes in those seeds are postcards from the Pleistocene, phone calls to the future of forestry. You can't buy them from Burpee's. So we gathered solid local support to save our last, best Willapa forest. Bert Cole—long-time commissioner of public lands, in the cockpit at DNR, one hand on the controls and the other on a chain saw—yielded to Brian Boyle, whose fresh set of policies emphasized the unique qualities of state forestlands. Among these, Boyle adopted a plan for a number of so-called "Old-Growth Seral Stage Deferrals." If granted, Hendrickson's deferral was to last until 1993; after that we hoped to dedicate this special place as a state natural-area preserve. In the end, though, only twenty-five acres of Hendrickson made it into this program.

We kept up the pressure. Professor Gordon Alcorn of the University of Puget Sound, eleven-year chair of the Scientific Area Advisory Committee, came to see the canyon with us and sent the commissioner his enthusiastic vote for protection as a Natural Area Preserve. The 1987 legislative session passed the Natural Resource Conservation Area Act, providing another means for removing ecologically significant trust lands from the timber harvest schedule. The next year the Wahkiakum County Commissioners, at our request, again wrote to Commissioner Boyle, asking that Hendrickson be considered as a Natural Resource Conservation Area. John Edwards replied for DNR that it was indeed a candidate.

A new commissioner, Jennifer Belcher, was elected on a conservation platform. She maintained the temporary administrative withdrawal for Hendrickson. But she was feeling pressure to make more state timber available for sale; formal preserve dedication would have to be garnered if we were to rest easy. At a one-on-one lobbying session with Commissioner Belcher at a function of the Nature Conservancy, I tripped and spilled a glass of red wine on her jacket just before her speech. The commissioner, fortunately, had a sense of humor and promised to send me her dry-cleaning bill. Late in 1994, she wrote to tell me that her office was "currently putting together a legislative packet to obtain the necessary funding for trust land transfer" for Hendrickson. But just as the site gained the top position in line, the legislature approved no funds for the transfer program, and we went back to writing and lobbying for another five years.

Meanwhile, a kerfuffle arose over the natural-area preserves. Some local people in Grays Harbor County, upset over potential restriction of their hunting, objected loudly to a proposed preserve on Elk River. Their howls alerted some politicians who introduced legislation to abolish or trim back the program. It failed, but the episode managed to put the state heritage protection effort on the defensive. Once more I asked the Wahkiakum County Board of Commissioners to pass a new resolution in favor of a Hendrickson Canyon reserve. Led by Esther Gregg and including Ron Ozment and Richard Marsyla, the county commission again came through with key support. So did the Gray's River Grange, of which I am a long-time member. Finally our district senator, Majority Leader Sid Snyder, a Democrat from Long Beach, gave his vital endorsement. Early in 1999, the legislature not only passed the appropriation, but placed Hendrickson Canyon in the top "must acquire" category—largely because of the key local support we had garnered.

On a vile day, just before Thanksgiving in 1999, we attended one last hearing before a DNR fiscal officer in the Pacific County courthouse. Opponents bantered about how we might all be spending the holiday together if the storm got any worse. The mood hardened as testimony began. A Pacific County commissioner with feet in both logging and real estate, a diehard opponent of state preserves, railed against removal of any land from either the tax base or the timber supply. How absurdly tiny our request was, I pointed out, compared to the hundreds of thousands of acres in the industrial forest estate, and how petty the argument against it. Fisheries manager Ed Maxwell spoke of the key importance of protecting the few remaining undamaged headwaters, when millions were being spent to restore damaged salmon streams all over the region. If it is cut, he said, it will be gone forever. Trees will come back, but not the forest, not in our time. When a Columbia tugboat crewman, Dan Toelkes, got up and shared from the heart his hopes that this one bit of woods might be spared as a token of what this place once was, the examiner had no doubt about the quality of local sentiment. The winterdark room, just then, felt like victory. But one hurdle remained, and it was a high one.

The outcome still had to be approved by the Natural Resources Board in Olympia. The board meeting came more than a year later, in early December. With it, a new and unexpected problem arose. Because the canyon is home to nesting marbled murrelets, a threatened species that came into prominence after the spotted owl, state appraisers rated its price tag very low. This meant the university trust would receive little value for the land and timber. There was a real danger of the Board's turning down the transfer on that basis. In-house economists had devised a bold scheme to apply substantial economic value for non-commercial habitat—a real advance in DNR thinking that promised to take the agency discourse beyond bare stumpage prices—but this new initiative threatened old-fashioned thinkers. The outcome would prove critical, not only for Hendrickson but for several other trust land parcels that were on the agenda as well, including old-growth additions to the Nemah Natural Resource Conservation Area and Willapa Divide Natural Area Preserve, located respectively west and east of our site. I'd been warned there might be disagreement, and I had almost stayed away from this meeting; I didn't think I could stand another let-down.

One member, the dean of the University of Washington College of Forest Resources, who turned out to have a remarkably brief tenure in that position, was wavering and seemed to be leaning against the initiative on the faulty theory that the land might be worth more later. The same Pacific County commissioner was furious that he had not been allowed to speak at this late date. Jennifer Belcher was chairing the Forest Board for the last time; after eight years in a job guaranteed to piss almost everyone off, she was not amused by the waffling and whining. Showing that she held no grudge over the red wine, she muscled the vote through in one of her last acts as Commissioner of Public Lands. On December 5, 2000, after a campaign of more than twenty years, Hendrickson Canyon finally was saved.

A celebration followed at the Fishbowl Brewpub in Olympia. Most of our Willapa old-growth campaigners came along. So did Laura Smith and Jennie Lange of the Nature Conservancy, an organization that had been key in getting the trust lands package through the legislature and the agency. Jack Scharbach joined us too. When Jack had first told me about Hendrickson Canyon, he had little hope of its protection; after all, every stand of trees in the Willapa Hills seemed bound to come before the saw one day. Now the deal was done. We lifted pints to the forest, to our friends, and to persistence, warm in the knowledge that at least in this small corner, the forest of our homeland would stand.

Long Canyon

Jerry Pavia

My relationship with Long Canyon, an eighteen-mile-long, unroaded, unlogged drainage in the northernmost part of Idaho, is filled with memories. When I look back over the years, all the efforts spent with politicians and conservationists to save Long Canyon from being logged come to mind.

In my file cabinet I have a set of old maps that show where the roads would have been built to enter Long Canyon and where the initial cutting units would have been placed to begin the logging of the area. Today, due to the continuous effort of conservationists, Long Canyon is no longer within the timber base and there are no plans to log the area. The maps are a curiosity now.

Looking in the mirror these days, I see a balding fellow with a white beard. So, this is what it looks like for me to be fifty years old. When I sit quietly and close my eyes, I can see pages and pages of writing about Long Canyon, but I can only share a small part. I am a skipping stone on a lake's surface as it touches here and there on its path to being eventually submerged in the lake itself. Notice if you can the ghosts crowding these pages, crying out to be included. Many will have to wait for a future time.

1973

All my climbing and backpacking gear is in the backseat. I am twenty-five. I have a full head of hair with a reddish gold beard. I am tanned from days spent on glaciers climbing mountains. There is no place I truly call home. I'm on my way from Oregon to Alberta to climb a large peak in the Canadian Rockies. It is summer. My climbing partner is asleep next to me as I drive farther north in Idaho towards the Canadian border. Sunrise. Pink clouds grace the sky. I come around a curve and am greeted by a spectacular vista of a valley with a meandering river bordered on the west by a mountain range. I try to wake my friend, but he only grunts and falls back asleep. I continue driving north. Shortly before crossing the border into Canada, I pull the car over and get out.

Looking across the valley to the mountain range I see the narrow mouth of Long Canyon. At that moment, I remember thinking that it would be fun to explore it. I moved to this place, started calling it home, and became involved.

Interlude

Up until the early 1970s, there was never a threat to log Long Canyon because of its narrow mouth. Building a road into the canyon from the valley floor would have been difficult if not impossible. But finally a road was punched into a neighboring drainage where it would be fairly easy to drop a road over the ridge into the canyon. The last major unroaded drainage in the Idaho Panhandle lay in danger. I imagine the forest of Long Canyon with its old-growth cedar, hemlocks, sub-alpine firs, larch, and white pine in some way sending out a call for help. If we do not try on new possibilities, we will become stale in our limited vision. By 1977, a group of activists is working to save the area.

1979

It is the ninth day of our stay in Long Canyon. We are camped twelve miles from the mouth of the canyon and the valley floor. We have been working on trail maintenance. People have come and gone, depending on how much time they could give. I have spent all nine days in the canyon.

I will be spending several more with my friend, Will, as we plan on hiking cross-country up to Smith Lake, a few thousand feet above the canyon floor. On this last morning it is raining harder than I've ever seen it rain in north Idaho. The canyon is not pretty. The trees, ominous. Low cloud cover. I begin to feel trapped.

As I hike down the canyon with Will, hardly able to see anything through the pouring rain, I have second thoughts about continuing with him up to Smith Lake. We stop. We are at the point where we need to start cross-country uphill. I tell him I'm not going. I can't go. I'm scared. I'm crying. Tears mix with the rain that is pounding my face, and together they flow through my beard and drop to the ground. He tells me that it's okay and heads off by himself. I continue down the canyon through rain and tears.

1982

With Will again. It's a day off from working on the trail, and we are exploring. It's what we do together. After several hours of looking at trees and moss and lichen and exploring the stream by jumping from boulder to boulder, we are on our way back to camp for dinner. On the trail, heading up the canyon to a campsite that is called "The Hilton," and right next to the trail we come across a bunch of blue chanterelle mushrooms in perfect condition for eating.

Just as Will gets ready to pick them, I yell stop. It is only then that I remember I am a photographer and have my medium-format camera in my daypack. I photograph the mushrooms. We return to camp where Will sautés them and shares them all around. Delicious! Several months later my photo of those chanterelles appears in *Audubon* magazine.

1985

Sitting with Pat by the campfire. She is the Forest Service trail-maintenance person with whom we work every summer in the canyon. It is late. No one else is up. We've been sharing things that are going on in our lives, for this is where we truly connect each year. It is a quiet night, the only sound that of sparks bursting into the air from the fire. The long-drawn howl of a lone wolf breaks the quiet. Pat and I look at each other.

Interlude

There are trees in Long Canyon that I always visit, a circle of seven white pines. Each of them is probably over two hundred feet tall with no limbs for the first eighty feet or so. In the dark of the night I slip out of my tent and stand naked in the midst of this almost-perfect circle of white pines. I look skyward. I can move upward with my being through these tall sentinels and explore the universe beyond.

1987

It is the sixth day of a solo cross-country hike I am doing along Lion's Head Ridge, Smith Ridge, and then down into Long Canyon to "The Hilton" and the twelve-mile hike out along the trail. I left Smith Lake hiking cross-country about an hour or so ago, heading for the canyon floor and the trail. I look around and see no one. I'm scared. I'm shaking. I've just fallen thirty feet down a cliff. My glasses flew off, but I found them unbroken. My legs are okay. My wrist hurts, and I later find out that it is broken.

For the first five days of the hike along the two ridges, my awareness was clear. The beauty of the sunsets. Stunted trees. Odd rock formations. A curious marmot that had probably never seen a human. From my camp in the evenings I howled at the stars and listened for a response. I wore a permanent smile. And just like that, on what I thought would be the easiest day of the hike, I wasn't paying attention, got myself into a place I shouldn't have been, and I fell. I learned that day that misplaced awareness can get me into trouble quickly. Awareness is a gift. Once I arrive safely home and have seen a doctor, I wear a cast on my arm for seven weeks.

1991

It's been a long, hot day. Dutch and I have spent it in a hole trying to dig one huge boulder out of the path of the redesigned trail. Dutch is sweaty, dirty, smelly and happy. So am I. We have been joking about spending several hours in the same place working on a rock. At the end of the day we finally succeed. One boulder removed. One big hole to fill with dirt. We now call ourselves the "Burly Brothers." There is no rock too big for us to move.

1994

On our way back from a hard day working on the trail, I ask Anne and Nils (an exchange student from Sweden) if they'd like to go talk to a tree. Nils looks confused. Anne looks curious. They both follow me up the trail. I tell them that I know a cedar that is at least eight hundred years old and that I'd like to introduce them. This is one of the trees that I always visit when I'm in the canyon. We go off trail a short distance, and there faithfully right in front of us is a magnificent western red cedar. We go up to it, and the three of us try to circle it with our arms outstretched. We fall far short. And then I start yelling at the top of my lungs at the tree: *You Are So Beautiful, I Love Trees*, and *Know That I was Here*. Anne and Nils have perplexed looks on their faces. I explain my belief that, because some can live for hundreds of years, you have to yell wonderful things to trees when you visit so that they will know you were there. After a short discussion about my reasoning, we are all yelling things to this old cedar.

1999

Sitting with Pat again, talking. We tell canyon stories. We each remember things that the other has forgotten. Do you remember? And then the laughter rings off the trees and moss and boulders and everything around us. Ghosts are sitting with us, pleased to be remembered. We are on a log over the creek that is next to "The Hilton." We have the canyon to ourselves. The others who worked on the trail left yesterday. We are spending one more day in the canyon just to relax.

My tent is nearby. I've camped in the spot for at least seventy nights over the years. My body knows the contours of this piece of ground. After Pat heads into camp, I jump from the log and go down to the water's edge. There is a still pool in front of me. Looking in, I see the reflection of both my face and the face of Long Canyon. We are inseparable.

Turnaround for a Threatened Wilderness

Karen Tweedy-Holmes

A tiny, desolate, and utterly surreal portion of northwestern New Mexico, a badlands area called Bisti/De-Na-Zin, powerfully attracts me. I photograph its contours, fantastic hoodoos, waterless washes, and changeable skies as a way to ground myself in the earthiness obscured by city living. Once a year, in this and other wilderness areas, I startle awake my awareness of a deep, ancient, and more exciting way of relating to the planet than I can find in a subway train, skyscraper, or even museum. The desert exhibits a monumental indifference to the concerns of our species—one of its paramount attractions.

In this country, we are rich enough to choose not to alter, deform, or rob every particle of landscape for economic gain. We rely on wild places, I believe, even if we never directly experience them. When development recently threatened the Bisti, I tried to use my photographs as totems against the threat of oil wells, pipelines, and new roads. Even if they don't have the power to stop this sort of destruction, the photographs can still bear visual witness.

I visited the Four Corners area of the Southwest for a few weeks every year for eleven years before I finally saw the Bisti. In the past, this forty-five-thousand-acre malpais held dinosaurs, crocodiles, turtles, and a wide range of tropical flora. Forests, swamps, and vast inland seas have alternately covered it. Today this irregular fifteen-mile stretch of trail-less desert receives an average of only nine inches of rainfall a year. But it is enough to support a diversity of wildlife: a few small rodents and lizards, beetles, spiders, nesting owls, and rare ferruginous hawks. These animals live only a short distance from the harm of human enterprise. When I visited the Bisti most recently, the nearby Navajo tribe was running seventeen "animal units," government-speak for a cow and her nursing calf, that scrounged for scant and edible green stuff in this arid place.

This unforgiving landscape has never been hospitable to human life. A few remains of seasonal camps from seventy-five hundred to eleven thousand years ago have been found among the dinosaur bones and fossilized trees scattered on the desert floor. Now only insect buzz, infrequent birdsong, and wind interrupt the silence. Over the eons, sandstone, shale, and coal have eroded into extraordinary shapes. Many of the sandstone hoodoos, in their sexual flowing forms, contain rhythmic black patterns that are the compact remnants of ancient plant debris. Even the sun's midday bleaching cannot flatten the profound colors of the rock and soil—the black, buff brown, sandstone gray, mudstone, ironstone, and brilliant crimson shale. The short, crumbly hills of shale formed when coal fires burned deep in the earth and cooked the iron-rich clay. Later the forces of erosion exposed and shaped these hills. Coal-black topsoil sets off shale mosaics of many reddish hues.

There are other surprises. Stromatolites, sedimentary stone ovals reminiscent of turtle shells, attest to the abundant microorganisms that lived in the ancient seas. A side-splotched lizard regenerating its lost tail climbs over the indentations left by water that flowed in the wash. When fog rolls in, usually in the autumn, the hoodoo edges soften.

✦ ✦ ✦

In 1991, Stephen Speer, a geologist from Roswell, New Mexico, purchased two leases totaling fifteen hundred acres from the Bureau of Land Management (BLM), the agency charged with the care of the Bisti. He wanted to drill for gas and oil. The abundant coal deposits of the area were too poor in quality to exploit commercially; attempts to mine them ended in the 1980s. Congress consolidated the Bisti and the neighboring De-Na-Zin Wilderness with a linking corridor in 1996.

Just two years later, Speer applied for a permit to drill on his leased land. He planned five gas wells and eight oil wells, to be serviced by more than eight miles of access roads. Prior to the Bisti's designation as Wilderness, the Navajo Nation owned the surface where the wells were to be drilled, and the BLM owned the subsurface minerals. Although the BLM had wanted to acquire the surface area in trade with the tribe as additional wilderness, such a trade seemed doubtful, so the agency felt that it had no justification for refusing Speer the leases.

When Speer sought to drill, the law required the BLM to produce an environmental-impact statement. Unflinchingly, the statement described the effects the proposed drilling and road building would have on topography, wildlife habitat, plant life, air quality, silence, and the aesthetic aspects that draw people to the region. After the statement appeared, the BLM held a series of public hearings and invited written response. The public opposed Speer's plans unequivocally.

Before the Bisti could be damaged, I visited the area in the autumn of 1999 to document it photographically. A friend accompanied me to help tote the amenities for a few days' stay. Our destination among the hoodoos was about two-and-a-half miles from the outer border of the wilderness. We packed in camping supplies, water, and photographic equipment.

We had fully expected blasting heat, but our time there was an anomaly —three days of intermittent rain. This weather was pictorial good fortune, the colors of the rocks vividly supersaturated, the meager flora full in leaf and bloom. Red-and-black-striped blister beetles scooted around in the moisture. When the sun emerged, curled arabesques of dried mud lined the drying washes. In the morning, though, the clay soil was treacherous. Two of my cameras and I hit rock when I took a tumble. Fortunately, all my equipment still worked.

My companion, a powerfully beautiful woman who was turning forty, was my other great good fortune. She was interested in discovering what complementary reflections her body might make with this most peculiar landscape. We spent one of our days making images of her nude form in contact with the innumerable weird shapes that wind and water had carved from rock. In a fair portion of the resulting black-and-white pictures, her body is nearly indecipherable from her surroundings. In others, her hair and limbs echo the hoodoos' shapes and invite the eye to dance to the rhythms of time. Later, when we viewed the images in detail, we were awed that the human and wild rock forms were so alike, so harmonic. This bodily connection intensified our desire to use the pictures to honor this landscape by drawing attention to its endangerment.

From our base camp we traveled to the area where the drilling was planned. Other visitors had been there before us. Of the original thirteen well pads that Speer had installed, twelve had vanished, sabotaged. For its unseemly contrast with its background, I photographed the thirteenth.

Drilling for oil in this area would have cost Speer a fortune, and strong public protest was ongoing. Speer offered to relinquish his right to drill for ten million of the U.S. taxpayers' dollars. The BLM declined the offer. Speer reduced his asking price to two million dollars; again he was rebuffed. Finally, he accepted $110,000, about $10,000 more than he had paid for the leases. The BLM assumed responsibility for plugging the single well hole and ripping out the road that led to it. The leases, reverting to the BLM, will never be renewed. The Bisti is safe from development for now, and if we take care to move lightly upon it, we may save it from other sorts of wounding.

Herons, Eagles, People, and Parks: A Cautionary Tale from a British Columbia Gulf Island

Rebecca Raglon

The water slapped and sucked against the embankment of the lagoon, and moving with the waves were the long thin legs of the dead heron. Its head rolled to one side, a golden eye open to the sky. For years I had watched the heron as it stood for long moments in the lagoon, poised like something from an exquisite Chinese silk painting. It was a bird whose vertical presence completed the sweep of the lagoon, the shiny mud flats, the logs stretched out across the water. And now, after a severe storm, the lagoon, the bay, the entire gulf island where I lived, seemed bereft without the heron: his harsh raucous call, his careful wading through the tides, his purposeful waiting.

In any case, the death of the heron went unnoticed and unmarked by any public ceremony. We were in the midst of a municipal election, and to everyone's surprise, it had turned into a rather rancorous and controversial affair when it was learned that the Greater Vancouver Regional District (GVRD) planned to sell off portions of Crippen Park as a cost-saving measure. Crippen Park was a five-hundred-acre park situated at the very entrance to the island where I live, right where the ferry from Vancouver docks. A long, straight road divides the stores, post office, and restaurants of the village on the left side of the road from the firs, alders, and blackberry bushes on the right. The park had been established years earlier and its origin was part of island mythology. The land had been slated for development with grandiose plans to install four thousand residential units, a golf course, and a ski hill. The Caterpillar to be used for land clearing had been brought over on a barge and landed on the beach with the precision of a military operation. At that time there were only eight hundred island residents—

hippies, eccentrics, protesters—but they had linked arms together against this onslaught and the outcome of the confrontation was the eventual establishment of the park by the GVRD.

Almost thirty years had passed, and how different things were now. With our past successes, we no longer felt a need to struggle. We were convinced that the worst excesses could be avoided. We believed in the environmental assessment process. We were sure others were paying attention and taking care of things for us. We were lulled by an environmental newspeak, which claimed that things were under control, that everything would work out in the end, and that we were all really good, green citizens—developers included.

Knowing that plans to sell portions of the park would be controversial, the GVRD hired a consultant who held a series of charettes (mock trial planning sessions) with invited members of community groups. The exercise that followed was a classic, ripped from the pages of a community planning textbook. As we walked through the park, we learned that the GVRD was going to "give" a chunk of land to the Arts Council for a performing arts center, another chunk to the recreation commission for playing fields, another chunk to the senior citizens for a senior citizen home, and on and on, with each community group lining up for some new goody. On the top of the list was a newly discovered community need for a large cement skateboard park.

In return, the GVRD wanted only a few acres of "surplus" land to be taken from the park and sold at market value. Together, environmental consultants and planners argued that the surplus land was mainly covered with weed trees—mature alders—with no real economic, aesthetic, or environmental value. Suddenly, everyone seemed to be supporting the dismantling of the park. If we didn't, how could we ever afford a home for our senior citizens? How could we ever afford a recreational center? Playing fields for children? Those few who wanted to save a few acres of weed trees were called tree-huggers and accused of being anti-people. Community leaders darted angry looks at each other at public meetings where the issue was discussed, and nasty, unforgivable things were said.

In the midst of these rancorous, ongoing negotiations, a pair of great blue herons established a nest in the park's "surplus lands"—right where the new community center was supposed to be located. I was delighted to see the herons back on the island. Their large, showy presence, the fact that they were nesting, their harsh calls, could hardly be ignored.

The fact that they were a vulnerable species also demanded our concern. One night a planner from Vancouver appeared on the island. She was dressed in a power suit, and she had brought a PowerPoint presentation that promised to address the issue of the herons. There in front of our eyes was a dazzling new picture of Bowen Island, complete with a "vibrant, bustling pedestrian mall" in the midst of the park. As for the herons, the planner pointed out with her little laser light, the *parking lot* outside of the community center would act as a buffer zone for the heronry. Community leaders—the local doctor, the baseball coach, the head of the arts council—almost fell over themselves in their eagerness to endorse this new plan. They described it as a win-win situation.

The election came and went. The incumbent, who had favored selling off surplus lands, was booted out, and a new green mayor was elected by the narrowest of margins. New, local committees were struck to study and delay the whole situation, and new plans were drafted. As humans continued to plan, however, the herons continued to settle among the weed trees of the surplus park lands: in three years the heronry has grown from one pair to eleven nesting pairs of herons. Twenty-two herons now perch on the trees above the ferry, wade in the lagoon, and circle in the sky above our village.

As the herons settled in, a group of twenty people formed a "Heron Watch" to study the birds during their nesting season. Led by Sue-Ellen Fast, a naturalist who also founded the Bowen Island Conservancy, the group split the days up into two-hour shifts to watch the impact of human activity—people coming off the ferry in their cars, car doors slamming, the sound of chain saws, lawnmowers, or hammers. The fact that herons could settle in what is a fairly busy area speaks to the fact that herons all across southern British Columbia are in a "vulnerable" state. Ironically enough, conservation success in another area—the protection of bald eagles—

has also contributed to the heron's decline as the eagles prey on herons. In June 2001, a well-established heronry on another Canadian gulf island was attacked by eagles. Over four hundred heron nestlings and eggs were destroyed, virtually wiping out the largest heronry on the west coast of Canada.

Ms. Fast believes that the herons located their nests in a busy, noisy area on Bowen Island because it offers a kind of "protection" from the eagles, which are reluctant to linger where there is a great deal of human activity. Over the three years of her research, Ms. Fast concluded that the herons are remarkably adaptable. They seemed able to distinguish between the regular noises in the village—the ferry, cars, lawn mowers—and unusual sounds emanating from within the park itself. In a presentation to the town council she spoke of the successes (one pair successfully raised three chicks) and failures (two chicks killed by bald eagles) and asked council to approve a series of guidelines to help islanders and herons get along.

What interests me in the story of the herons is that, as people bickered and quarreled over a patch of weed trees, the herons arrived, and settled in, and by their actions made all other arguments moot. Any space, it seems, even those deemed "not important" by paid environmental consultants, may be of crucial importance to another species. Further, the struggle on my island is similar to what is happening all across British Columbia, all down the coast, all across the West. Each struggle is a crucial, vital struggle, even though there are those who will tell you that weed trees aren't important or that land somewhere can be sacrificed for a "greater good" somewhere else. The ability to "dig in" or "draw the line" in towns and neighborhoods everywhere might at times seem trivial or hardly worth the effort when compared to saving old growth forests or grizzly bears, but in reality, it is a crucial part of any conservation strategy. Each struggle—for a patch of trees, for a path leading to a beach, for stream enhancement, for neighborhood parks, for a bog, a marsh, a patch of grass—is part of a greater mosaic. It is a mosaic created as each community struggles in its own way to preserve the living integrity of its home. This is our job, and it's tiring and time consuming, and it doesn't have the adrenaline rush of past struggles, when protestors sat in front of an approaching Caterpillar. We live in new times, and the work that needs to be done is often at town council meetings and committee meetings. We scrutinize bylaws and zoning amendments, we keep track of covenants and setbacks, each of us drawing comfort from one another. We know that in every town and village, in every neighborhood, there are others alert enough and concerned enough to do the same and, in the process, to make a real difference.

Chaparral Gold: Point Mugu

Bill Weiler

Los Angeles is not all traffic and smog. Forty miles up the coast rises the crown jewel of the Santa Monica Mountains: Point Mugu State Park, thirteen thousand acres of chaparral green and sycamore sienna. It is truly the last remaining intact coastal canyon of significance in the entire Santa Monica Mountains. The other canyons are known as Los Angles suburbs. No Hollywood glitter here, yet plenty of gold. Mugu's paragon of beauty, La Jolla Valley, contains a species of native bunchgrass not found anywhere else in the entire mountain range, as well as many prehistoric Chumash Indian cultural sites. Yet if you believe that areas located in the state park system are permanently protected, pull up a chair, and I'll weave a true story of two teenage boys and their special place.

In 1969, Governor Ronald Reagan approved a disastrous plan to drastically change the focus of state park management in California. In the past the policy had been to leave parks in their semi-natural state with limited development—trails, picnic tables, and campground areas. The new recreation policy, however, emphasized expanding: resorts, hotels, and restaurants, as well as a gas station, gift shop, archery facility, food market, variety store, swimming center for two thousand plus a two-thousand-car parking garage straddling a future freeway, an eighteen-hole golf course and clubhouse, and more.

Point Mugu, though originally purchased exclusively to be a primitive park, was to be Governor Reagan's first showcase of his new policy. His man for the job: William Penn Mott, Jr., State Parks director at the time (later the National Parks director under President Reagan). I was fifteen, and the weekend had come. My buddy Walter and I spent nearly every weekend bike riding the forty-five miles to Point Mugu Recreation Area. We were smitten by the grace of the tree-lined canyons, wind-chimed Sycamore Creek, and the sublime views and peacefulness of La Jolla Valley. On one unforgettable Friday evening, we arrived at dark. A group meeting was in progress and, noticing the profusion of environmental bumper stickers on their vehicles, we realized quickly that here was a party we should crash. To our horror, the chatter centered on the terrible, unfathomable development scheduled for our beloved park. Like patriotic nationalists, we immediately enlisted for the frontline. Walt and I decided that if we had to, we would hide in the park and perform acts of civil disobedience, even if it meant skipping homework for a few nights. Actually, little did I know that the next two years of my life would be fully immersed in a war that seemed impossible to win. So many classes would I miss that my mother had to plead permission from the principal for her son to graduate.

The next day as Walt and I pensively walked down the familiar canyon trail, we schemed about our own new plan, the formation of an environmental organization specifically committed to protecting Point Mugu. It was a perfect time to form the Save Point Mugu Committee, even if it consisted only of two charter members. Our best attribute was our boundless energy, our worst, a lack of any experience in politics and organizing supporters. We would use the school print shop to produce custom bumper stickers and flyers. We were determined not to let school studies interfere. Five months later, I would run into my estranged English teacher, who chided, "It would be nice to see you in class some time."

The first political spectacle I ever experienced unveiled itself at the first Point Mugu Management Plan public hearing in Oxnard, California. In the spotlight beamed the confident sixty-one-year-old, white-haired park veteran, William Penn Mott Jr. Mott's claim to infamy: in the 1930s, working for the National Park Service, he researched the feasibility of establishing the Lake Tahoe region of Nevada and California as a national park. Mott's report sadly recommended against park status. Here are a few actual Mott quotes from the seven-hour public meeting:

> On the rationale: "I'm planning parks for people not for birds, animals, and shrubs."
>
> On the swimming area: "Take those cold rapidly moving currents of a mountain stream. Youngsters raised in cities are often startled

by them and would prefer to swim in a large pool just like those back home."

On the whole: "After all, other state park systems and the national parks follow us, we don't follow them."

Only one person, representing motorcycle interests, spoke in favor of the proposal. The other six dozen speakers, botanists, archaeologists, politicians, and local residents, made their thumbs-down-to-development views clear. Sitting next to me, a new high school partner drew Gary Larson-type caricatures of the state park officials, while performing a voodoo curse on them.

Two weeks later, we were thrown for a loop. Mott hit the press proudly announcing that as a result of the heavy opposition to the original development, the plan had been "severely cut back." The rifle range, model-airplane area, golf course, tennis court, and hotel/motel complex had all been axed, but the bulk of the mini-city scheduled for La Jolla Valley was still intact.

The State Parks Department contended that three-fourths of the twenty thousand expected Point Mugu visitors would now be recreating on the park's sandy beach instead of in La Jolla Valley, even though the uplands were the area where all the fun-filled activities were to be located. As far as my buddy and I were concerned, the remaining commercial entities still proposed for Mugu were too many, too much. When destroying a natural area, there is little difference between commercialization and over-commercialization. We geared up for the next public forum: the California State Parks and Recreation Commission meeting to take place one week later in Palm Springs.

In Palm Springs, we felt like political pros. We packed the hall with a raucous standing-room-only crowd. The commission members were an interesting bunch, including the ex-football-player chair, whose opening remarks sounded quite reasonable. Unfortunately, five of his fellow commission members acted like they wanted to go home before the first gavel sounded. One of them insisted that all parks be opened to hunting. Another member, after hearing enough opposition to Mott's remarks, asked if any of us wanted to take his position on the commission, "then please come forward." The throng lunged to the front. One of the two open-minded members asked those in the audience who opposed the plan to stand up. Virtually everyone did. "What more evidence do you need?" he asked his fellow commissioners. At the hearing's end, Mott asked the commissioners to delay their decision on Mugu till their next meeting to be held one month later in Oroville, four hundred miles north of Los Angeles and Point Mugu.

Walt and I dug deep into our pockets and were able to buy plane tickets to Sacramento. Our plan: hitchhike ninety miles from Sacramento to a campground just a few miles from central Oroville, then somehow hitchhike to town in time for the next morning's meeting. Walt and I found ourselves in Oroville by sunset, and both knew that the campground we were searching for waited only a few miles out of town. We were willing to walk there with our backpacks if no one picked us up. No one did, and a cold February night soon laughed at us. "Just around the next bend" became the running joke as Walt's watch pointed at ten-thirty; we had been walking down the highway for nearly five hours. A car packed with peers sped by us, but magically stopped one hundred yards downwind and turned around.

"What are you two fools doing at this hour?" they mocked.

"We're headed toward the campground for the night. How far is it?"

Shaking his head, the driver laughed. "If you mean the state park, another eleven miles. . . . Hop in, we'll give you a lift."

Next morning, the hitchhiking was much easier. Thanks to two key rides, we arrived in Oroville at exactly 9:00 A.M., the start of the meeting. We became depressed when we noticed the football chairman was conspicuously absent. He was our swing vote, and now we wondered if thugs from the parks department had not kidnapped him. The commission immediately voted to disallow public testimony. Mott opened the meeting with a few surprises. He recommended eliminating the motorcycle campground, clubhouse restaurant, and inn. That announcement prompted cheers from the dozen of us who made up the Los Angeles contingent. It is difficult to describe the rest of the proceedings. One commissioner presented his fellow members a box with 1,106 letters received during the last four weeks. All but six opposed the state's vision for Point Mugu. He also stated that the plan would still ruin 250 acres of the 400 native bunchgrass acres in La Jolla Valley. No matter, the commission approved the mass "wreck-reation" project by a five-to-two vote, in addition to a terrible amendment calling for Point Mugu to be reclassified as a state park. The amendment would have been a good thing, except that the commission also rescinded its own state park policy, leaving no official guidelines as to what could or could not be included in a state park. Walt and I were so incensed we sent smoke signals to each other, the plumes being emitted from our ears.

Our battlefront was expanding. We now attacked the lifeline to the Mugu plan, its funding, by actively lobbying the California State Legislature. Sixteen million dollars was needed, but had not yet been approved for sewer, water lines and road construction. Without those funds, the project was doomed. Walt and I lobbied everyone we could to send letters urging the acquisition of natural areas adjacent to Point Mugu, including Bony Mountain and other tracts, which would double the size of the park. In the meantime, to Walt's and my great surprise, the Save Point Mugu Committee had received close to fifteen dollars in contributions. Our bumper stickers and mailing expenses came to only ten, so we spent the rest on ice cream.

Months later, Walt and I skipped school again, returning to Sacramento to testify before the sub-body of the State Assembly Ways and Means Committee. Both the Assembly and the Senate Finance Committee were examining four thousand bills and bond projects, which included Point Mugu, determining whether or not to fund them. The House Public Works board chairman was moved by our slide show and was sympathetic to our cause. He wondered how our high school education could possibly compare with the experience of the Mugu campaign. We also visited with Willy Brown (now mayor of San Francisco), who buoyed our spirits by saying that the governor's budget request would have to go through him, and he was not impressed.

However, the Senate Finance Committee chairman was a cement-the-world fellow. Nothing was certain. We needed more contributions in order to pay for our airplane tickets. Despite months of planning, lobbying, and long-distance public testimony, Mugu's final fate rested out of our realm. Our last hope, the state legislature, convened behind closed doors.

Walt and I and another buddy couldn't stand the tension so we decided if we were going to be thoroughly familiar with the park, we'd have to camp in its backcountry. Now we knew that everyone was required to stay at the drive-in Sycamore Canyon Campground, but we were low-impact campers and needed to find secret caves to hide in, if ever we launched commando raids on survey stakes and bulldozers. As night replaced day, we found a nice spot at the far end of Sycamore Canyon, but some fuddy-duddy must have tipped the rangers off, for no sooner were the sleeping bags unfurled than the state park ranger slowly drove his truck up the trail/road with spotlights combing the bushes. We gave a team hush and the light beam failed to unveil us. The next day, we decided to hike out through La Jolla Valley, but those rangers were smart—one of them met us as we were just leaving the park. He was stern, yet friendly, but best of all, he brought news: The Senate Finance Committee had just stripped the Point Mugu development fund from Governor Reagan's proposed budget! The plug had been pulled. The two of us were ecstatic, laughing and crying while hugging each other and even the ranger. A couple of amateur fifteen-year-olds pitted against the world, and we won!

One year later, the Parks and Recreation Commission designated a large portion of La Jolla Valley as a natural preserve. Further into the seventies, thousands of neighboring acres would become the 150,000-acre Santa Monica Mountains National Recreation Area, though only one-fifth of that amount had been purchased by 1982.

I returned recently to be reacquainted with the winds at La Jolla Valley. Upon arriving at dawn, I knew where I was and my long-sought reunion was worth the wait. The light sienna of the wild grass was perfectly accented by the dark green chaparral behind it. My eyes gazed in four directions. Invisible, thank Mott, was the walk-in campground he had managed to sneak in. Wherever it was, the camp was well hidden, the only sign being a distant small patch of manicured green grass. Oh, oh. A facsimile of the dreaded water line had come. God forbid that hikers should be required to carry water into the valley on their own backs!

Old memories, like a bright candle, had been rekindled with the advent of a new day's light. I was gladly immersed in the present. Here together—the mountain lion, the wind, and I joined hands for a reaffirming instant, then went our separate ways.

In View of the Condor

Bradley John Monsma

After hiking for nine hours, my friend and I crest a ridge alongside the Sespe Condor Sanctuary sixty miles northwest of downtown Los Angeles. To the south and west, darkening silhouettes of the Channel Islands brood as lights flicker on in the beach cities. L. A. begins to glow in the southeast. To the north, the wilderness, I trace the flow of Sespe Creek, the last major undammed stream in Southern California. Four thousand feet above the creek, steep bluffs in the sanctuary redden then fade. In my mind's eye I see all the views at once; portions of landscape I usually experience separately begin to connect. As we catch our breath and savor the scenery, my friend breaks the reverie to remark, "If condors can't survive here, they deserve to die. What more can we do for them?"

I chuckle at his candor. It's true this looks about as good as it's going to get from here on out for condors. A wilderness at our feet, we seem to be at the top of the known world, on the verge of taking flight ourselves. We can feel the updraft. If I were alone I would at least stretch my arms and hop a little to see if I would have the faith for an enskyment. What more could a bird, even one with a nine-foot wingspan, need? But the more I try to see the land as a condor might, the more I begin to think differently about the history of this place. I begin to see that the lines we've drawn to protect condors reflect human more than avian perspectives. Consider two perspectives on distance. My friend and I began our hike by crossing the middle of the sanctuary at its narrowest point, about three miles by flight, a few more by trail. By the end of our three-day, thirty-mile walk and the car shuttle, we had circumnavigated the western two-thirds of the sanctuary. By contrast, condors can glide at fifty-five miles per hour and range three hundred miles in a day on the lookout for dead deer, cattle, ground squirrels, or defrosting calves left by Fish and Wildlife. The same space that seems nearly endless measured foot by plodding foot might appear relatively small in the eyes of a condor on the wing. The Sespe Condor Sanctuary is only part of the terrain condors need for a full life.

I've tried to see the Sanctuary and the rest of the Sespe backcountry as a condor might by sitting on the high ridge where I thought of taking flight and glancing between the terrain and a map. Only on the map can I see the distinct lines separating the Sespe Condor Sanctuary from the Sespe Wilderness and the official wilderness from the rest of the Los Padres National Forest. Apart from the map, the terrain arranges itself into brushy creek beds, hillsides shifting between chaparral scrub and forest, sandstone outcroppings, and occasional meadows. This is what I imagine a condor might see. Only for a short period have condors and this place been separated— between 1987, when the last wild condor was captured, and 1992, when birds hatched and raised in captivity began to be released into the wild. The rest of the time, condors have ignored the lines we've drawn over the last century. From our perspective, though, some of the boundaries owe their existence to the birds.

To understand this, we need to go back to 1939, when pioneering condor researcher Carl Koford began to portray condors as extremely shy and sensitive to intrusion. His descriptions became conventional wisdom and served as the basis of an argument about the best way to forestall extinction. Since condor numbers were declining and human intrusion might be to blame, it stood to reason that the best way to save condors was to preserve habitat and thereby keep people away from the birds. According to Noel and Helen Snyder, authors of the most recent and comprehensive book on the California Condor, Koford's notes and correspondence indicate that the idea of a sanctuary was there from the beginning of his research, and in 1947, the Sespe Condor Sanctuary was established.

Later, in the 1980s, land preservationists working to designate what was then called the Sespe-Frazier Wilderness quickly recognized that if habitat might save condors, condors might also save habitat. A charismatic, endangered bird might sway public opinion and become a symbol of the need to protect wild lands. The symbolism was crucial since the struggle to preserve habitat was occurring simultaneously with the controversy about whether to remove birds from the wild to begin a captive-breeding program. Opponents of the program argued that captive breeding a wilderness icon would be a step towards a time when all wildness would be regulated, monitored, and controlled. Wilderness activists feared that once condors were absent from proposed wilderness areas, the general public would see no reason to protect their habitat. The famed wilderness advocate David Brower, whose organization Friends of the Earth opposed capturing wild condors, described the connection between the Sespe and condors with his typically pithy phrasing, "A condor is 5 percent feathers, flesh, blood, and bone. All the rest is *place*."

Given the persistent threats to the sanctuary and its surrounding wilderness, the use of the condor as symbol seems quite understandable. The region has been the site of oil and natural gas exploration since the late 1880s, and at least five proposals to dam Sespe Creek have failed or been defeated. Despite both wilderness designation and Wild and Scenic River status for Sespe Creek being granted in 1992, state legislators still call for dams on the Sespe to solve the latest energy crisis.

In hindsight, however, ironies abound. Those who argued most forcefully against removal and captive breeding suggested that scientific research on condors was incomplete and that in our ignorance we had best preserve land and let condors do their thing as they always had. It seems now that their position was also based upon incomplete scientific knowledge of condor mortality, especially of the role of lead poisoning and power line collisions. It would seem that even had the greatest possible area been given the highest degree of protection, condors might still have become extinct without captive breeding. It is entirely possible that had opponents of the condor program prevailed, the condor may have made the dignified exit some suggested was the best course in bad times. On the other hand, their arguments signaled a shift in conservation from emphasizing single species to focusing on ecosystems and biodiversity. Due in part to the efforts of these activists and the condor's charisma, we now have protected wildernesses an hour's drive from massive cities in a growing state. These largely intact ecosystems preserved with condors in mind remain the geography of hope for other endangered species—southern steelhead, red-legged frogs and arroyo toads, bighorn sheep—as well as many less charismatic, less threatened creatures.

Condor habitat, however, and condors on the wing, cross even the boundaries of what we consider ecosystems and watersheds. While the sanctuary and wilderness areas are important for nesting and roosting, they offer very little foraging habitat. Historically, condors have flown outside the protected areas to find food, particularly to private ranches where for hundreds of years they have feasted on dead cattle. Now, these ranch lands are rapidly turning into housing developments as Los Angeles's suburbs spawn suburbs of their own.

Still, as I write in the spring of 2001, condors are nesting in Southern California. They lay eggs and raise chicks in the wild for the first time since birds hatched in captivity have been released. The condors are intensely managed, but their presence, even compromised by radio collars and supplemental feeding, reminds us of the vitality of the places we've saved for them. For me, this is true even when I don't see them, which is most of the time. A regular feature of my backcountry Sespe walking is the fleeting hope conjured by a shape in motion in the corner of my eye. "It's a bird, it's a plane, it's...." A biologist with the Condor Recovery Program told me of a "Condor Eye Chart" someone put up in the Fish and Wildlife Office. A profile of a condor takes the place of the big "E," and as the chart descends into rows of smaller figures, condors mix with hawks, eagles, airplanes, butterflies, and bumblebees. The possibility of seeing a condor helps me know the place in all its complexity.

When I have seen condors, usually aided by a biologist with a telemetry unit, my view of them—and through them—alters my sense of this place. A pair might appear as specks first against mountain then sky, rising on a thermal in wide, fourteen-second turns, marked by their unique steadiness in flight. Then, at freeway speed, they approach, ducking behind a wrinkle in the landscape. Out of view for a few moments, the habitually curious birds appear suddenly overhead where they slow to gaze down, their brilliant white underwings flashing. Too soon, they turn and match the falling angle of the landscape before catching an updraft. As I watch the pair, specks again heading south, joy and thanksgiving filter through the gray-brown haze of Los Angeles in the background. It occurs to me that the condors can look from downtown L.A. toward the heart of the Sespe with a turn of their pinkish, featherless heads. Their long flights remind me of the meagerness of what we've left them. Their view holds no illusions about the separateness or sufficiency of what we label habitat, wilderness, or sanctuary. Attempting to see the lay of the land through their eyes becomes the act of discipline that conditions my appreciation of the Sespe.

Amazing Grace

Kathleen Dean Moore

Three miles downstream from the covered bridge on the Mary's River, there used to be a dam. The Mary's is a small river west of the Cascade Mountains, barely a creek by Oregon standards. It was not much of a dam either, just a three-foot-high concrete wall built by a farmer to power a paddle wheel for his private electrical plant. By the time my husband, Frank, and I bought the land on both sides of the river, the farmer had moved on, the pastures had grown up to Queen Anne's lace and bracken, the paddle-wheel blades had rusted off and washed downriver, but the dam remained. We decided to blow it up.

Even though it was insignificant as dams go, we worried that our dam made life hard for Willamette River cutthroat trout. In late winter, when the water is still high, the cutthroats migrate from the Willamette into the feeder streams of the Mary's River watershed to spawn. But when the Mary's starts to warm up in late spring and water levels drop, many of them head downstream again to spend summer in the deeper river. Our dam blocked the Mary's at low water, forcing the fish to hang out in slackwater, waiting for rain. The dam blocked canoeists, too, who had to portage through a wicked blackberry bramble or jump out of their boat, haul it across the dam, and climb in again—a tricky maneuver.

Taking out a dam wasn't a political issue ten years ago, and it never occurred to us that it might be a legal issue or a decision controlled by layered federal agencies. Somebody put a dam in; somebody can take it out. As Frank and I talked it over, we made no fine distinctions among blowing up, taking out, breaching, and notching a dam. We just wanted to get the dam out of the way, and dynamite was good enough. We made arrangements indirectly, through the man who sold us the land.

A fist of concrete in the center of the road was the first sign of trouble when we drove from home to the farm the day after the dam came out.

Even before the river came into view, we could see rocks littering the roof of the neighbor's barn, and when we climbed through blackberries to the creek, we found rubble scattered bank to bank. Twisted rebar stuck out of the river, each post catching a little vee of twigs and dried leaves. The guy must have drilled some holes, stuck in some dynamite, lit a fuse, and run like hell.

We apologized repeatedly to the neighbors, but they were just grateful that by some blind miracle we hadn't brained their cow. Wading into the river, we started sawing out the rebar with the neighbor's hacksaw. It was a wet business, leaning over at the waist, up to our armpits in silty water, sawing away. The river was so muddy that we stumbled around, feeling with our feet for rubble, stopping sometimes to wiggle our tennis shoes loose from the silt. All afternoon, we hauled blocks of concrete out of the riverbed, stacking them as artfully as we could along the edge. Every time I stood to stretch my back, I heard restless water and felt the awakening river push against my legs.

As hard as we worked to clean the bed of the little reservoir, the river worked harder, lifting clouds of silt from the gravel and swirling them toward a beach at the bend downriver. By the end of the day, what had been slackwater was a stream again. Exhausted, we sat on the roots of a willow tree and watched the river, washed silver by a watery sun emerging between clouds and hills. I wish I could say there were trout leaping in rainbow arcs over the river. I wish I could say there were trumpets. There were not. But from every rush and backwash, each swell of water over rock, each small spurt and riffle, each surge, all the slosh and spill of a river on the move, came a music we had never heard in that place.

Now, ten years after the great demolishment, what was a silty slackwater river bottom is scrubbed to bedrock, and moss and iris crowd the rocky edge. The river runs strong across a riffle, hesitates, then dives in a smooth swell over the sill where the dam used to be. The Mary's still has its problems—agricultural runoff from pastures upstream and clearcutting on the highest hills. But cutthroat trout breed freely in the little streams upriver and, on a good day, the water runs clear under the roots of the willows.

And now, all across the country, people are taking out dams and restoring free-flowing rivers. Kennebec, Naugatuck, Chipola, Bear, Walla Walla, Whitestone, Souadabswok—almost five hundred dams from California to Maine. In the Pacific Northwest, my corner of the continent, everybody's talking about the dams in the Columbia Basin. Maybe engineers will draw down the reservoir behind the John Day Dam on the Columbia River, uncovering miles of cobble riverbed that someday might be full again of salmons' arching backs and flashing tails and streamers of bright red eggs. Or maybe engineers will breach the four dams in the lower Snake River, and when the reservoirs have drained away and floods have scoured the river, find spawning beds and smooth rock walls where there was only mud. Then, the salmon could return—thousands of salmon, millions of salmon, silver shapes leaping up the whole lively length of the Columbia River.

The excitement comes not only from activists and fishery biologists and politicians but from friends and families and our neighbors in the little town we live in. The idea captures people, raises their pulse. It's wild, hopeful talk. "Maybe it's not too late," says the man down the street. "It could happen. In my lifetime. Imagine." He smiles a long smile and folds his hands behind his neck.

Last week I drove out of the wet valleys of western Oregon and turned east along the length of the stair-stepped reservoirs that Northwesterners still affectionately call the Columbia River. On the east side of the Cascade Mountains, this is scabland, a steppe of sage and bunchgrass prairies, layer on layer of basalt laid down by lava flows and shaped by floods into brown velvet hills, greening in the creases on this spring day, capped with fraying clouds and scalloped with rimrock and high-tension wires from hydroelectric plants. Drifts of purple lupine were just beginning to flower. At the place where the river once turned on edge and thundered between basalt cliffs, where native people stood on platforms and hauled salmon from the maelstrom in long-handled nets, where in March 1957 the floodgates of the Bonneville Dam closed and a reservoir drowned the river overnight,

I stood on the lawn by the water and tossed pieces of my McBreakfast to a gull. The noise was deafening—roaring trucks on the highway, bells at the railroad crossing, and then the rapid pulse of an empty train. The river itself was silent.

I have come to believe that dam-breaching is not really about dams. It's probably not even about fish. Dam-breaching is America's own exercise in truth and reconciliation. For a hundred years, we thought we could have it all—cheap power, salmon, and alfalfa fields in the desert—but we were wrong. We thought we could capture the great rivers in the West, put them to narrow human purposes, and pay no price. Wrong again. We thought it made perfectly good sense to transport salmon in trucks on the highway, so that grain and petroleum could move in barges on the river. We thought we needed power and wealth, but we discovered to our sorrow that what we really need are health and beauty and a way of life that listens to the land. In the Columbia Basin, once home to ten million wild salmon, endemic Snake River coho salmon are extinct. Snake River chinook are endangered. Sockeyes are threatened.

What humans destroy, we often destroy forever. When lumbermen cut an ancient forest, another like it will not grow in my lifetime or my grand-children's, not in fifteen generations. When the last member of a species dies in a zoo, it is gone forever. But a river? A river has the power to forgive. To breach a dam is to admit mistakes, to release the power of the river to heal itself, to begin to heal the rift between human and nature, user and used.

Sitting in my car behind a chain-link fence that separates the highway from the river, I can imagine people lined up at the edge of the Columbia, watching as water drains through the John Day Dam and the riverbed rises slowly into that blue desert air. Water pours from every crease, and smooth basalt slabs appear, gray and mammoth. People will pull into the highway overlooks, stand in small groups under cottonwoods in the riverside parks, wade across the mud to pile stone cairns at the falling edge of water—crowds of people, gathering quietly to witness the rebirth of a river.

In the first days, all they will hear is a whisper, the movement of silt in warm water. But as the lake falls away and a rocky island splits the channel, the river will shout over rapids and then begin to roar, a full-throated roar lifting the screams of gulls and the laughter of children who jump from rock to rock, leaving their footprints in soft sand beside the prints of geese, scooping minnows from pools of water stranded inside rotting tires.

    Maybe the people will cheer. Maybe they will pray. Maybe they will weep when they see the pale riverbed, drowned for a very long time. But the first rain will clean the highest rocks, the first flood will cut a channel through the silt. Storksbill and balsamroot will poke up between slabs of mud, drying on new riverbanks, and I know from experience that there will come a time—maybe a very long time, but in our lifetimes if we live right—when the roots of willows will reach into clear water again.

Notes on Contributors

David Axelrod is a professor of English at Eastern Oregon University. He received the 2003 Kay Deeter Award for his poetry. His essays and poems have appeared in *Kenyon Review, Cimarron Review, Boulevard, Quarterly West*, and *From Here We Speak: An Anthology of Oregon Poetry*. He is the author most recently of a collection of nonfiction, *Troubled Intimacies: A Life in the Interior West* (Oregon State University, 2004), and poetry, *The Chronicles of the Withering State*. His newest collection of poems, *The Cartographer's Melancholy*, will be published by EWU Press in Fall 2005.

Michael P. Branch is a professor of literature and environment and Fitzgerald Distinguished Professor of the Humanities at the University of Nevada, Reno. He is a co-founder and past president of the Association for the Study of Literature and Environment, book-review editor of the journal *ISLE: Interdisciplinary Studies in Literature and Environment*, and co-editor of the University of Virginia Press book series *Under the Sign of Nature: Explorations in Ecocriticism*. His recent books include the Pulitzer-Prize-nominated *John Muir's Last Journey: South to the Amazon and East to Africa* (Island, 2001); *The ISLE Reader: Ecocriticism, 1993–2003* (co-edited with Scott Slovic, University of Georgia, 2003); and *Reading the Roots: American Nature Writing Before Walden* (University of Georgia, 2004).

Dan Brister is the project and communications coordinator with the Buffalo Field Campaign. He holds an M.S. in environmental studies from the University of Montana and a B.A. in English from the University of Vermont. His essay is drawn from a book-length manuscript, *In the Presence of Buffalo*. To help protect the Yellowstone herd, write to dan@wildbison.org.

Colin Chisholm lives in Missoula, Montana. His work has appeared in numerous magazines and anthologies, including *Audubon, Utne Reader*, and *High Country News*. His essay "A Place Worth Fighting For" was a finalist for the National Magazine Award. His first book, *Through Yup'ik Eyes: An*

Adopted Son Explores the Landscape of Family, was published in 2000 by Alaska Northwest Books.

Laird Christensen's unruly résumé ranges from lumber grader to environmental activist, park ranger to professor. His poems and essays have appeared in a variety of journals, including *Utne Reader*, *Wild Earth*, *Northwest Review*, and *Whole Terrain*. He currently chairs the department of English and communications at Green Mountain College, Vermont's environmental liberal arts college.

Christine Colasurdo is a poet and journalist, and the author of *Return to Spirit Lake: Journey Through a Lost Landscape* (Sasquatch, 1997) and *Golden Gate National Parks: A Photographic Journey* (Golden Gate Parks Association, 2002). Her work has appeared in *Audubon*, *Orion*, *California Wild*, and other publications. She teaches calligraphy in San Francisco and volunteers as a member of the San Francisco Green Schoolyard Alliance at her son's elementary school, where weedy dirt is being transformed into gardens for kids and wildlife.

Terrell Dixon teaches literature and the environment at the University of Houston. His essays on both wilderness and urban nature have appeared in numerous books and journals, and he co-edited *Being in the World: An Environmental Reader for Writers* (Longman, 1992). The University of Georgia Press published his *City Wilds: Essays and Stories about Urban Nature* in 2001.

Jim Dwyer is the bibliographic services librarian at California State University, Chico. He is the author of *Earth Works: Recommended Fiction and Nonfiction about Nature and the Environment* and is currently writing a guide to ecofiction. His performance poetry nom de plume is the Reverend Junkyard Moondog.

Bruce D. Eilerts is statewide manager of natural resources for the Arizona Department of Transportation. Previously he worked as associate director for the Center for Biological Diversity and as director of the U.S. Air Force's

Barry M. Goldwater Training Range in Arizona. He specializes in wildlife biology, natural resources management, and environmental planning and compliance.

T. Louise Freeman-Toole's memoir about life in Hells Canyon, *Standing Up to the Rock* (University of Nebraska, 2001), won both the Idaho Book Award and the Pacific Northwest Booksellers Award in 2002. Her work has appeared in numerous journals, newspapers, and anthologies. She has received fellowships from the Howard Foundation, the MacDowell Colony, and the Vermont Studio Center and is currently a Steinbeck Fellow at the Center for Steinbeck Studies in San Jose, California.

Harold Fromm is visiting scholar in English at the University of Arizona, co-editor of *The Ecocriticism Reader: Landmarks in Literary Ecology* (University of Georgia, 1996), and a regular contributor to the *Hudson Review*. His most recent work has been on Darwinism in the humanities.

Amanda Gordon teaches writing in the English department at Whatcom Community College and is a family literacy educator at the Center for Children and Families. She has her master's degree in English and lives in Bellingham, Washington.

Estar Holmes, a reporter with the St. Maries *Gazette-Record* in Idaho, covers the Coeur d'Alene Indian Reservation. A graduate of Evergreen State College in Olympia, Washington, she coordinated the grassroots group Dawn Watch. Estar has written for regional newspapers and has had pieces published in *Earth First! Journal* and *Higher Truth* magazine.

William Johnson teaches at Lewis-Clark State College in Lewiston, Idaho. He is a Northwest native and has published a critical study, *What Thoreau Said: "Walden" and the Unsayable* (University of Idaho, 1991), and two collections of poetry, *At the Wilderness Boundary* (Confluence, 1996) and *Out of the Ruins* (Confluence, 2000), the latter the winner of the Idaho Book of the Year Award. From 1998 to 2001 he served as Idaho State Writer in Residence.

Derrick Knowles works for the Northwest Ecosystem Alliance in Spokane, Washington, to protect northeastern Washington's remaining wilderness. He holds an M.A. in technical communication.

Carolyn Kremers is the author of *Place of the Pretend People: Gifts from a Yup'ik Eskimo Village* (Alaska Northwest, 1996). Her essays and poems have appeared in *American Nature Writing 1999*, *Brevity*, *Creative Nonfiction*, *Manoa*, *Newsday*, *North American Review*, and elsewhere. She has served on the M.F.A faculty at Eastern Washington University and currently teaches at the University of Alaska-Fairbanks.

Paul Lindholdt, an English professor at Eastern Washington University, also serves on the boards of the Northwest Fund for the Environment and the Upper Columbia River group of the Sierra Club. He has won awards from the Academy of American Poets and the Society of Professional Journalists, and has published more than 150 books, articles, reviews, essays, poems, and editorials about American culture and the environment.

Bradley John Monsma teaches literature, writing, and California natural history at Woodbury University in Burbank, California. He just published a cultural and natural history of Sespe Creek, the last free-flowing stream in Southern California, entitled *The Sespe Wild: Southern California's Last Free River* (University of Nevada, 2004). He is a passionate but prudent whitewater paddler and a backpacker trying to learn the flowers and go light.

Kathleen Dean Moore is the author of *Riverwalking: Reflections on Moving Water* (Harvest-Harcourt, 1996) and *Holdfast* (Lyons, 1999), winner of the 2000 Sigurd Olson Nature Writing Award. Her most recent book is *The Pine Island Paradox* (Milkweed, 2004). She is a professor of philosophy at Oregon State University, where she teaches philosophical field courses and directs the Spring Creek Project for Ideas, Nature, and the Written Word. Moore lives on the wet side of the Cascades in Corvallis, Oregon.

Jerry Pavia has lived in Bonners Ferry, Idaho, since 1975. He worked in the timber industry for eleven years before becoming a freelance photographer. He has been the sole or principal photographer for thirteen books, and his work appears regularly in national magazines. He chaired the Idaho Conservation League from 1996 through 2001, and he is currently on the board of the Yellowstone to Yukon Conservation Initiative.

Will Peterson founded Walrus and Carpenter Books in Pocatello, Idaho, in 1988 and the Rocky Mountain Writers Festival in 1990. He is the author of four books of poetry: *Luctare Pro Passione* (Walrus and Carpenter Books, 1995), *The Flows, Atlas of the Unknown West*, and *Love Poems from Mink Creek*.

Chuck Pezeshki is a writer, photographer, and environmental activist, and also an engineering professor at Washington State University. He is the author of *Wild to the Last: Environmental Conflict in the Clearwater Country* (Washington State University, 1998) and currently works on U.S. and Canadian wild forest protection issues.

Robert Michael Pyle dwells along a tributary of the lower Columbia River, where he writes essays, poetry, and fiction. He holds a Ph.D. in ecogeography from Yale University. His fourteen books include *Wintergreen* (Houghton Mifflin, 1996), winner of the John Burroughs Medal; *The Thunder Tree* (Houghton Mifflin, 1993); *Where Bigfoot Walks* (Mariner, 1997); *Chasing Monarchs* (Mariner, 2001); *Walking the High Ridge* (Milkweed Editions, 2000); and several widely used butterfly books, including the recently published *Butterflies of Cascadia* (Seattle Audubon Society, 2002).

Rebecca Raglon worked as a tree planter, telephone operator, and newspaper reporter in many parts of the Canadian bush (the Yukon, the Northwest Territories, and northern British Columbia), before returning to school to complete her degrees. She lives on Bowen Island, British Columbia, and teaches at the University of British Columbia.

Sharman Apt Russell lives in New Mexico. Her books include *An Obsession with Butterflies* (Perseus, 2003); a collection of botanical essays, *Anatomy of a Rose* (Perseus, 2001); a novel on Paleolithic life, *The Last Matriarch* (University of New Mexico, 2000); *When the Land Was Young: Reflections on American Archaeology* (University of Nebraska, 2001); and *Kill the Cowboy:*

A Battle of Mythology in the New West (University of Nebraska, 2001). She teaches at Western New Mexico University and at Antioch University in Los Angeles.

Robert Schnelle is an instructor at Central Washington University. He has published an essay collection, *Valley Walking: Notes on the Land* (Washington State University, 1997), as well as articles in *Seattle Review*, *Weber Studies*, and *Writing on the Edge*. He lives with his family near Ellensburg, Washington.

Lee Schweninger is a professor of English at the University of North Carolina at Wilmington where he teaches courses in American Indian literatures and environmental literature. He has written several essays on American Indian writers and a book on N. Scott Momaday. He is currently at work on a collection of narratives by Colorado pioneer women and another book-length collection of personal essays. He also volunteers on the North Carolina Nature Conservancy's burn crew.

Sherry Simpson teaches creative nonfiction writing in the M.F.A program at the University of Alaska, Anchorage. She is the author of a collection of essays entitled *The Way Winter Comes* (Sasquatch, 1998), which won the inaugural Chinook Prize. Her essays have appeared in anthologies and magazines including *Creative Nonfiction*, *Sierra*, *American Nature Writing*, *Going Alone*, *Gifts of the Wild*, *On Nature: Great Writers on the Great Outdoors*, and others. She has won the Andres Berger Nonfiction Award from Northwest Writers, Inc.

Mitchell Thomashow is the chair of the Antioch New England department of environmental studies and founder of its doctoral program. His first book, *Ecological Identity: Becoming a Reflective Environmentalist* (MIT, 1995), offers an approach to teaching environmental education based on reflective practice. Thomashow's second book, *Bringing the Biosphere Home* (MIT, 2001), is a guide for learning how to perceive global environmental change. The founder of the periodical *Whole Terrain*, he serves on the advisory board of the Orion Society.

Stephen Trimble, a Salt Lake City naturalist, writer, and photographer, has won the Ansel Adams Award from the Sierra Club. His eighteen books include *The Sagebrush Ocean: A Natural History of the Great Basin* (University of Nevada, 1999); *The People: Indians of the American Southwest* (School of American Research Press, 1993); *The Geography of Childhood: Why Children Need Wild Places* (with Gary Nabhan), published in 1995 by Beacon; and *Testimony: Writers of the West Speak on Behalf of Utah Wilderness* (with Terry Tempest Williams), published in 1996 by Gibbs Smith.

Karen Tweedy-Holmes is a freelance photographer based in New York City, with an abiding passion for the deserts and canyons of the Colorado Plateau. Her photography of the Bisti/De-Na-Zin Wilderness, funded by the Mindlin Foundation, has been exhibited at the Population Council in New York City and at the Holden Art Center at Warren Wilson College in Swannanoa, North Carolina. Her most recent projects include documenting threatened areas of the Grand Staircase/Escalante region of Utah and hiking and photographing in the Grand Canyon. She is collaborating with Derrick Jensen on a book about the incarceration of animals. Her work appeared in the April 2003 issues of *The Sun* and *Heartstone* magazines.

Stacy Warren teaches geography at Eastern Washington University. She nurtured her interest in culture and landscape growing up in the Pacific Northwest and the Los Angeles area. She is the author of articles chronicling the geographic impact of the Disney Company, and her current research focuses on environmental change and postmodern culture in the land of smog, earthquakes, and theme parks.

Bill Weiler is a habitat biologist for the Washington Department of Fish and Wildlife and vice-president of the Central Cascades Alliance. He edited *The Earth Speaks* (Institute for Earth Education), a 1983 anthology by people who live in close contact with the natural world, and is finishing the book *Carrying a Blanket of Stars: Lessons in Developing One's Sense of Wonder.*